BINARY DIGITAL IMAGE PROCESSING
A Discrete Approach

BINARY DIGITAL IMAGE PROCESSING
A Discrete Approach

Stéphane Marchand-Maillet[⋆], Yazid M. Sharaiha[*]

⋆ *Department of Multimedia Communications*
Institut EURECOM
Sophia-Antipolis – France

* *Imperial College of Science, Technology and Medicine*
London SW7 – United Kingdom

San Diego San Francisco New York Boston
London Sydney Tokyo

This book is printed on acid-free paper

Academic Press
24–28 Oval Road, London NW1 7DX, UK
http://www.hbuk.com/ap/

Academic Press
A Harcourt Brace and Technology Company
525 B Street, Suite 1900, San Diego, California 92101–4495, USA
http://www.apnet.com

ISBN 0-12-470505-7

Library of Congress Catalog Card Number: 99-64631

A catalogue for this book is available from the British Library

Transferred to digital printing 2005
Printed and bound by Antony Rowe Ltd, Eastbourne

CONTENTS

LIST OF FIGURES

LIST OF ALGORITHMS

FOREWORD

Numerical representations of images are currently developed at different levels of perception with CCD cameras, computer processing and visualisation with numerical photography. Since the fundamental sampling theorem was developed from the field of information theory, new techniques have been widely integrated in digital image processing. Significant research results have been offered in the area of analysis, synthesis and communications. These results are underpinned by geometrical notions, which are always present for image support representation, surface characterisation, object description and shape modelling and within all the tools dedicated to visualisation, animation and object modelling.

An important amount of work has been devoted to transcribing continuous concepts arising from Euclidean geometry into digital spaces. A classical approach first consists of embedding discrete data into a continuous representation. A geometrical process can then be executed on such a continuous representation. Finally, a discretisation step returns to the digital representation.

Such a processing sequence of approximation techniques is based on interpolation models between the continuous and discrete spaces. By contrast, a non-classical approach consists of transcribing Euclidean geometry (definitions and properties) into the discrete space, and this is the domain of Discrete Geometry. The challenge is in the transformation approach and related descriptors. Does the discrete notion tend to its continuous counterpart when the discrete data tends to the continuous one? This is a very difficult problem for which significant research activities are engaged.

With this book, the authors illustrate the fundamental aspects of digital image processing using elementary examples. Starting from topological notions, geometrical ones are studied and illustrated via metrics, straight lines and convexity. Whenever possible, practical algorithms are described and illustrated in detail. An entire part is devoted to applied combinatorial mathematics and algorithmic graph theory. Basic notions are introduced and developed from arithmetic, graph-theoretic and combinatorial concepts, where the main effort of the authors is to combine the related domains in order to offer a relevant presentation and associated proofs for discrete geometry.

The authors are acknowledged for offering us this original contribution. The book should be of interest to a large group of researchers and engineers in computer science and applied mathematics. Such a book is also of fundamental support for lecturers, particularly with the increasing integration of image processing courses. It should bridge the gap between the fields of computational and differential geometry and algorithmic graph theory within the more general image processing literature.

Jean-Marc Chassery
Director of research at CNRS, France.

ACKNOWLEDGEMENTS

Most of this text was written while the authors were with Imperial College, London.

The authors thank Dr John E. Beasley, Imperial College, for reviewing this book and for his useful comments. The technical assistance of Dr Pascal Gros from Institut Eurécom in the production of this manuscript is acknowledged.

The first author wishes to make a special dedication to Bidel for her invaluable support and help during the process of writing this book.

To our parents

NOTATION

Notation is kept consistent throughout the book. Unless otherwise indicated, the following notation is implicitly used.

Sets

$\emptyset$	Empty set	
$\{p\}$	Set containing p as a unique element	
$\{p_i\}_{i=0,\ldots,n}$	Set of $n+1$ discrete elements p_i	
$\mathbb{N}$	Set of all positive integer numbers	
$\mathbb{N}^*$	Set of non-zero integer numbers	
$\mathbb{Z}$	Set of all integer numbers	
$\mathbb{Q}$	Set of all rational numbers	
$\mathbb{R}$	Set of all real numbers	
P	Set of discrete points or set of pixels	
F	Set of foreground pixels	Def. 1.12
F^{C}	Set of background pixels	Def. 1.12
V	Set of vertices in a graph	Def. 3.1
V_T	Set of vertices in a tree	Def. 3.10
A	Set of arcs in a graph	Dcf. 3.1
A_T	Set of arcs in a tree	Def. 3.10
$\overline{A}$	Arcs in A without their orientations	Rem. 3.2
S	Set of skeleton points	Prop. 7.1
F_n	Farey sequence of order n	Def. 1.40

Operations on sets

$S_1 \cup S_2$	Union of sets S_1 and S_2	
$S_1 \cap S_2$	Intersection of sets S_1 and S_2	
$S_1 \setminus S_2$	Set difference (S_1 "minus" S_2)	
S^{C}	Complementary set of the set S	
∂S	Border of the set S	
$\overline{S}$	Closure of the set S ($\overline{S} = S \cup \partial S$)	
$\mathring{S}$	Interior of the set S ($\mathring{S} = S \setminus \partial S$)	
$\|P\|$	Cardinality of the set P (number of elements in P)	
Γ	Border of a set of discrete points P	Def. 1.11
$[S]$	Real convex hull of the set S	Def. 2.27
$\langle S \rangle$	Lattice points contained within $[S]$	Def. 2.29

Values

∞	Infinity (a large number in practice)	
D	Discrete distance value	
n	Cardinality of an arc, path or set	Def. 1.5
ε	Threshold value or scale factor	
i, j	Generic integer numbers	
k, k'	Connectivities	Def. 1.11
a, b, c	Move lengths on the square grid	Def. 1.16
N, N_G, N_T	Number of vertices	Def. 3.1, 3.10
M, M_G, M_T	Number of arcs	Def. 3.1, 3.10
σ, μ	Slope and shift values of a real straight line	
x_α, y_α	Coordinates of point α	

Points and vertices

p, q	Pixels or discrete points (points with integer coordinates)	
α, β, γ	Real points (points with real coordinates)	
u, v	Vertices in the graph	

Functions

$i \leftarrow j$	Set i to the value of j	
$\lfloor x \rfloor$	Floor function	Def. 1.21
$\lceil x \rceil$	Ceiling function	Def. 1.21
$\mathrm{round}(x)$	Rounding function	Def. 1.21
$\lVert\cdot\rVert_2$	Euclidean norm	
$i \bmod j$	Modulo function	
$\gcd(i, j)$	Greatest common divisor between i and j	Def. 1.40
$\oplus$	Minkowski addition	Rem. 6.3
$\ominus$	Minkowski subtraction	Rem. 6.3
$\circledast$	Hit or miss operator	Def. 7.13
$d_{\mathrm{E}}(\alpha, \beta)$	Euclidean distance between α and β	Def. 1.25
d_{D}	Generic discrete distance	
$d_4(p, q)$	City-block distance between p and q	Def. 1.26
$d_8(p, q)$	Chessboard distance between p and q	Def. 1.28
$d_{a,b}(p, q)$	Chamfer distance between p and q	Def. 1.35
$\mathrm{DT}_{\mathrm{D}}(p)$	Distance transform based on d_{D} at point p	Def. 5.1
$\mathrm{EDT}(p)$	Euclidean distance transform at point p	Sect. 5.3
$\chi_4(S)$	Crossing number of set S	Def. 6.6
$\mathcal{C}_8(S)$	Connectivity number of set S	Def. 6.6

Other notation

$\forall \ldots \exists \ldots \in \ldots$	"for all" ... "there exist" ... "in" ...	
$[\alpha, \beta]$	Continuous closed segment from α to β (α and β are included in the segment)	
$[\alpha, \beta[$	Continuous semi-open segment from α to β (α only is included in the segment)	
$]\alpha, \beta]$	Continuous semi-open segment from α to β (β only is included in the segment)	
$]\alpha, \beta]$	Continuous open segment from α to β (α and β are excluded from the segment)	
$\{a_{ij}\}_{i=0,\ldots,n;i=,0\ldots,m}$	Matrix formed by $n \times m$ elements a_{ij}	
N_4, N_8	Neighbourhoods	Sect. 1.2
(u, v)	Arc between vertex u and vertex v	Def. 3.1
$l(u, v)$	Length of the arc (u, v)	Def. 3.5
$w(u, v)$	Weight of the arc (u, v)	Def. 3.5
$G = (V, A)$	Graph based on sets V and A	Def. 3.1
$\overline{G} = (V, \overline{A})$	Undirected version of $G = (V, A)$	Rem. 3.2
$T = (V_T, A_T)$	Tree based on sets V_T and A_T	Def. 3.10
P_{pq}	Digital arc between p and q	Def. 1.4
P_{uv}	Path between vertices u and v	Def. 3.7
$\Delta_k(p, r)$	Discrete disc of radius r centred at p for the k-connectivity relationship	Def. 1.20
$\Delta_\mathrm{E}(p, r)$	Euclidean disc of radius r centred at p	

PREFACE

Interest in automated image processing has grown steadily in the last few decades. Both theoretical and practical advances have been made, strongly related with those of computer science. A common representation of a two-dimensional digital image is an array of pixels (picture elements) associated with colours in a colour map. The particular case of digital images that can be represented using a two-colour palette (foreground and background) defines the class of binary digital images. In binary digital images, information is contained in the shape of components and the inter-relationships between them. More specifically, binary line images are the images in which the foreground components have a ribbon-like structure. In other words, the shape of these components is much greater in length than in width. Instances of this class naturally include handwriting, engineering drawings and road map images. Hence, the use of binary digital images is spread in a wide range of applications including image analysis and image recognition. A robust theoretical background has become of primary importance as the applications grow more sophisticated. Foundations of digital image processing have been based on various already mature theories such as information theory, theoretical computer science, and combinatorial and discrete mathematics. Advances in image processing have also been made in parallel with these areas. However, it is now clear that digital image processing represents an independent area of research in itself rather than a sub-domain of any other subject.

This book has specific goals. Firstly, it should represent an introduction for a beginner in the field. Basics of binary digital image processing are therefore assimilated in the first few chapters. This book should also be a reference document to the student, the researcher and the industry specialist in the domain of binary digital image processing. Finally, this book is also addressed to the specialist as it contains a summary and survey of some of the up-to-date techniques in the field. It has been our wish to maintain a comprehensive and consistent style when writing this book, detailing results with examples and constructing theories from generally well-known background. The book is composed of three parts, each representing an aspect level of binary digital image processing.

The three first chapters set the theoretical context in which image processing operations and algorithms will be defined.

Chapter 1 presents the construction of a consistent topology in the discrete space. This chapter provides robust basic definitions such as connectivity and distance. Comparisons in relation to the continuous space are given and detailed throughout the study.

Chapter 2 presents geometrical definitions and properties of discrete sets including straightness, convexity and curvature. These definitions and properties are mostly drawn from Euclidean geometry. Again, consistency is maintained with the continuous space and comparisons with equivalent continuous concepts are given.

Chapter 3 introduces and relates applied combinatorial mathematics and algorithmic graph theory in the context of digital image processing. New definitions on the analogies with digital image processing are given.

Building on this theoretical background, the three next chapters present operators that will be the basic tools used in image processing applications. At this stage, pixels forming images are grouped into sets that are characterised and treated as wholes.

Chapter 4 introduces the first step of any image processing application, namely, the acquisition of the image. The acquisition process is modelled as a mathematical operator, in order to control and quantify the quality of the approximation of continuous images by digital ones. Data structures used in this context are presented and compared.

Chapter 5 presents a fundamental image analysis tool, the Distance Transformation. The Distance Map resulting from this operation often forms the representation of the image on which the rest of the processing will be undertaken. Efficient algorithms for Distance Transformation, involving different approaches are presented and compared. The linkage between Euclidean and discrete Distance Transformation is studied in depth.

Chapter 6 introduces the basis of shape characterisation of an image by detailing the common characteristics considered in image processing. The relationships of such characteristics with noise are considered and studied. Definitions and algorithms are presented in this context.

Finally, the two last chapters present definitions of operators on pixel sets or image components. The global structure of the image is now introduced in the models.

Chapter 7 defines thinning (or skeletonisation) operations on binary digital images. The specific class of binary line images is formally defined and modelled in order to adapt thinning operators. This operation results in the skeleton of the image. Models for the skeleton are presented and different approaches for obtaining it are studied and analysed. In particular a graph-theoretic approach is presented which exploits the results from Chapter 3. This context is shown to be well-suited for binary line image analysis. The aim is to define invariant image characteristics on which high level analysis such as classification and recognition can be performed. An introduction to binary line image vectorisation based on the output of the thinning operation concludes this chapter.

Chapter 8 presents some practical applications of different classes of binary digital images. This chapter illustrates how the techniques presented throughout the book can be applied for performing image analysis. The use of the different approaches is discussed in conjunction with the specific applications proposed.

Chapter 1

DIGITAL TOPOLOGY

Owing to the discrete nature of computers on which automated image processing is to be performed, images are typically given as sets of discrete points. In order to obtain a robust mathematical background for digital image processing, a formal study of such sets is to be developed. Then only, theoretical investigations can be carried out for presenting digital image processing operators.

The aim of this chapter is therefore twofold. Firstly, the acquisition step whereby a continuous set is mapped onto a set of discrete points is briefly introduced in Section 1.1 in order to characterise discrete sets which represent binary digital images. Then, a topology is to be build in this context. Sections 1.2 and 1.3 address this problem from the basis of neighbourhood relationships to the definition of discrete sets. Based on the results derived in these sections, Section 1.4 presents the construction of discrete distance functions. Finally, Section 1.5 studies the compatibility of such distance functions with Euclidean distance. This last part also allows for the refinement of the definitions of discrete distance functions.

1.1 Continuous to discrete images

The acquisition of an image is generally done using a set of physical captors. The acquisition process can therefore be accurately modelled as a sampling of the continuous image using a discrete partitioning of the continuous plane. For the sake of simplicity, only partitions involving regular polygons are considered. That is, polygons with sides of constant length and a constant angle between them.

It is easy to show that, for constructing a partition of the plane, only three regular polygon types can be used. The possible numbers of sides of the regular polygon used are three, four and six, leading to triangular, square and hexagonal partitioning schemes, respectively (see Figure 1.1). Physically, such polygons represent captors sensitive to the intensity of light. Their output is a value on a scale. A polygon area of constant colour is called a picture element or pixel, for short. Colour captors output three (e.g. RGB) values per pixel whereas grey scale captors output only one (e.g. grey tone) value per pixel. In a grey scale image, each pixel is therefore associated with a single colour value.

In the mathematical model of an image, the pixel area is identified with its centre, leading to the representation of pixels as discrete points in the plane. As

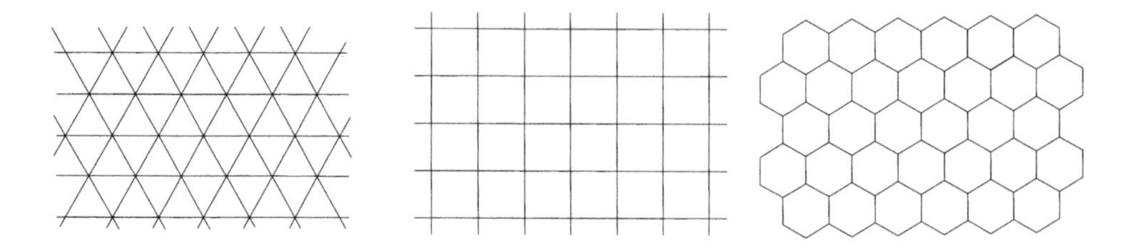

Figure 1.1 Different sampling schemes

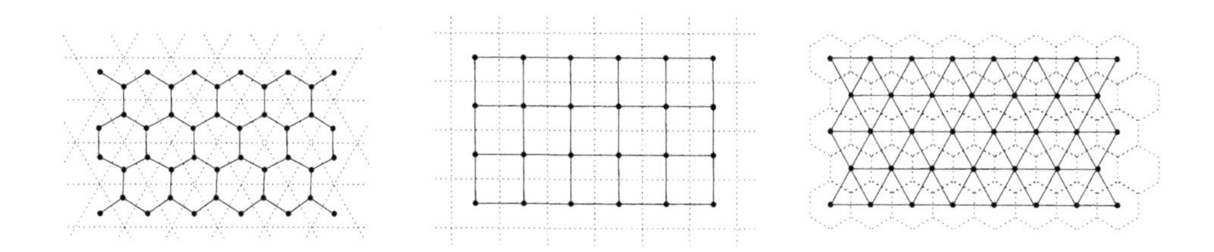

Figure 1.2 Resulting pixels from sampling shown in Figure 1.1

shown in Figure 1.2, a lattice can be built which connects all such pixel centres. The sampling partition is represented with dotted lines and the pixel centres as black dots ($\bullet$). The lattice represented with continuous lines is dual to the partition in the sense that two pixels are joined in the lattice if and only if the two partition polygons share a common edge.

A triangular partitioning results in an hexagonal lattice. Conversely, a hexagonal partition will result in a triangular arrangement of pixels, the triangular lattice. Finally, for a square sampling of the image, the pixels can be considered as points of a square lattice.

As in the sampling of the spatial domain of the image, the colour scale is sampled using a given number of discrete ranges (e.g. 24-bit colour images). We consider grey scale images where the colour scale is one-dimensional. When using only two such ranges representing white and black colours (0 and 1, respectively), we obtain binary images.

As result of the complete acquisition process, a two-dimensional binary image is given as a two-dimensional array of pixels where each pixel is associated with a colour value which can be either 0 (white pixel) or 1 (black pixel).

In order to define mathematical tools for picture processing such as connectivity and distance measurement, we need to construct a theoretical basis on the discrete set of pixels thus obtained. Digital image processing relies heavily on the definition of a topology which forms the context in which local processing operators will be defined.

Remark 1.1:

At this stage, digital images are studied as set of discrete points. However, the problem of mapping a continuous image onto a binary digital image will be studied in depth in Chapter 4.

From now on, the binary image is represented by a set of discrete points lying on a regular lattice. With each point is associated a 0-1 value which indicates the colour of its corresponding (white or black) pixel. We lay down a theoretical context for the study of digital image processing operators by first defining a topology on this set of points. At this stage, the lattice is considered as infinite (rather than finite) in order to avoid dealing with specific cases that arise close to the border of a finite lattice.

1.2 Neighbourhoods

It is commonly known that the discrete topology defined by pure mathematics cannot be used for digital image processing since in its definition every discrete point (i.e. a pixel in the image processing context) is seen as an open set. Using this definition, a discrete operator would only consider the image as a set of disjoint pixels, whereas in image processing, the information contained in the image is stored in the underlying pixel structure and the neighbourhood relations between pixels.

Alternative definitions have been proposed. In contrast with classic *discrete topology*, digital image processing is based on *digital topology* [21, 64, 138]. The definition for digital topology is based on a neighbourhood for every point.

Neighbourhoods in digital topology are typically defined by referring to the partition dual to the lattice considered. For a given point, defining its neighbouring points is equivalent to defining a relationship between the corresponding pixel areas in the partition. The simplest instance is when the neighbours of a pixel are defined as the pixels whose areas share a common edge with the pixel area in question (direct neighbours). Extensions for this principle are also considered. In this section, the three possible regular lattices (triangular, hexagonal and square) are investigated. For each case, commonly used neighbourhoods are presented. Section 1.3 completes these definitions by deriving properties on the neighbourhoods.

1.2.1 Triangular lattices

In the particular case of a triangular lattice, the neighbours of a point are defined as the six direct neighbours of this point on the lattice. By duality, it is equivalent to consider neighbouring (hexagonal) pixel areas as the ones which have a common edge with the pixel area in question. This neighbourhood is referred to as 6-neighbourhood on the triangular lattice. The notation $N_6(p)$ will be used for the 6-neighbourhood of point p. In Figure 1.3, the point p is linked to its 6-neighbours by bold lines. The triangular lattice is represented with thin lines. The dotted lines show the pixel areas dual to the triangular lattice.

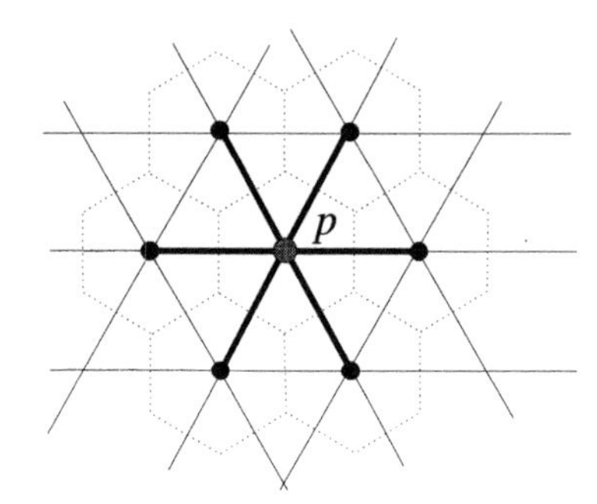

Figure 1.3 $N_6(p)$: 6-neighbourhood of the point p on the triangular lattice

1.2.2 Hexagonal lattices

A similar definition to the one above for the hexagonal lattice leads to the 3-neighbours of p on the hexagonal lattice (see Figure 1.4(A)). However, in Section 1.3, it will be shown that this neighbourhood is too coarse to satisfy basic properties in digital topology. Therefore, we extend this to the 12-neighbourhood as follows.

The 3-neighbours of a point p are defined as the points which are associated with the pixel areas that share a common edge (i.e. a one-dimensional object) with the pixel area in question. Following this principle, nine extra neighbours can be defined as the pixel areas that share a common corner (i.e. a zero-dimensional object) with the central pixel area (indirect neighbours). Combining, we obtain the 12-neighbourhood of the point p. By analogy with the previous section, $N_3(p)$ and $N_{12}(p)$ will denote the 3- and 12-neighbourhoods of a point p, respectively.

Figure 1.4(B) illustrates the 12-neighbourhood of the point p. The (triangular) partition is also shown as dotted lines to illustrate relations between the corresponding pixel areas.

Remark 1.2:

In contrast to the 3- and 12-neighbourhoods, we can note that the 6-neighbourhood defined in the Section 1.2.1 readily contains such an extension. In other words, on triangular lattices, there is no possible definition for indirect neighbours. Figure 1.3 shows that all the possible connections between the central hexagonal area and its neighbours are exploited.

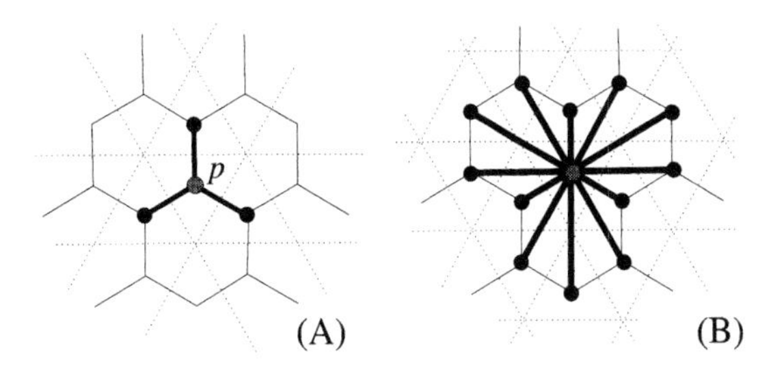

Figure 1.4 Neighbourhoods on the hexagonal grid. (A) 3-neighbourhood, $N_3(p)$. (B) 12-neighbourhood, $N_{12}(p)$

1.2.3 Square lattices

Four main neighbourhoods are generally defined on the square lattice. Firstly, the 4-neighbourhood ($N_4(p)$) includes the four direct neighbours of the point p in question (see Figure 1.5(A)). By duality, they are the pixel areas which share a common edge with the centre pixel area. This neighbourhood is completed using pixel areas which share a common corner with the pixel area in question (indirect neighbours), leading to the 8-neighbourhood of the point p, $N_8(p)$ (see Figure 1.5(B)).

By analogy with a chess board, the 8-neighbourhood corresponds to all possible moves of the king from p on the chess board. Extending this analogy, the *knight*-neighbourhood ($N_{\text{knight}}(p)$) which corresponds to all possible moves of a knight from p can also be defined (see Figure 1.5(C)). Finally, the combination of the 8- and the *knight*-neighbourhoods, yields the 16-neighbourhood of p, $N_{16}(p)$ (see Figure 1.5(D)).

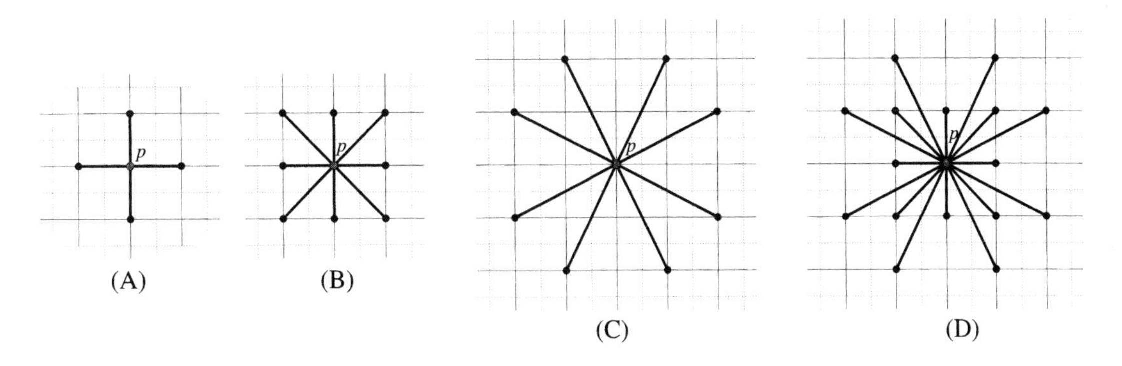

Figure 1.5 Neighbourhoods on the square grid. (A) $N_4(p)$: 4-neighbourhood. (B) $N_8(p)$: 8-neighbourhood. (C) $N_{\text{knight}}(p)$: *knight*-neighbourhood. (D) $N_{16}(p)$: 16-neighbourhood

Remark 1.3:

It is important to note that the square lattice is simply a translated version of its dual partition. Moreover, the position of the points on this lattice is well suited for matrix storage. For these reasons, neighbourhoods on the square lattice are the most well studied and the most commonly used. Rosenfeld [138] defined digital topology on this lattice.

1.3 Discrete sets

The definition of neighbourhood allows for the definition of local connectivity between points. Digital arcs and curves are simply an extension of this property. In turn, they impose conditions on their underlying neighbourhoods.

1.3.1 Digital arcs and closed curves

Definition 1.4: Digital arc

Given a set of discrete points with their neighbourhood relationship, a digital arc P_{pq} *from the point* p *to the point* q *is defined as a set of points* $P_{pq} = \{p_i \; ; \; i = 0, \ldots, n\}$ *such that:*

(i) $p_0 = p$, $p_n = q$.

(ii) $\forall \; i = 1, \ldots, n-1$, p_i *has exactly two neighbours in the arc* P_{pq}, *the points* p_{i-1} *and* p_{i+1}.

(iii) p_0 *(respectively* p_n*) has exactly one neighbour in the arc* P_{pq}, *namely, point* p_1 *(respectively* p_{n-1}*).*

Definition 1.5: Cardinality of a digital arc

n *is called the cardinality of the digital arc* P_{pq} *and is also denoted* $|P_{pq}|$.

A set of points may satisfy the conditions for being a digital arc when using a specific neighbourhood but the set may not satisfy these conditions for a different neighbourhood. Since most definitions and properties depend on the neighbourhood used, we specify this dependence by adding neighbourhood prefixes (i.e. 3-, 4-, 6-, 12- or *knight-*) to the names of the properties or digital objects cited. For instance, a digital arc in the 6-neighbourhood on the triangular lattice will be referred to as a 6-arc. Equivalently, a 6-arc is a digital arc with respect to the 6-connectivity relationship.

Using the definition of a digital arc, a connected component on the lattice is defined as follows.

Definition 1.6: Connected component

A connected component on the lattice is a set of points such that there exists an arc joining any pair of points in the set.

A further restriction on connectedness leads to definition of a simple connected component (and simple connectivity).

Definition 1.7: Bounded and simple connected component

On the infinite lattice, a connected component that contains an infinite number of points is said to be unbounded. On the finite lattice, a connected component is unbounded if and only if it intersects the border of the lattice. Otherwise, it is said to be bounded.

A simple connected component is a connected component whose complement does not contain any bounded connected component.

By definition a digital arc is a simple connected component.

An important notion in the continuous space is that of closed curves which, in turn, defines holes. In the continuous space, Jordan's theorem characterises a closed curve as a curve which partitions the space into two subparts, the interior and the exterior. The definition of a closed curve in the discrete space relies on that of a digital arc.

Definition 1.8: Digital closed curve

A digital closed curve (or equivalently, a digital curve) on the lattice is a set of points such that the removal of one of its points transforms it into a digital arc.

A version of Jordan's theorem in the digital space can be then formulated.

Theorem 1.9: Discrete Jordan's theorem

A digital curve defines exactly two separate connected components on the lattice, the interior and the exterior. Therefore, there should be no arc joining these two subsets.

Remark 1.10:

Theorem 1.9 emphasises the fact that, by definition, a digital closed curve is not a simple connected component since its interior is bounded (i.e. contains a finite number of points).

This theorem strongly depends on the notion of connectivity and, therefore on the definition of the neighbourhood used for each point. Hence, it is important to detail the cases corresponding to each possible neighbourhood used. This study finalises that of neighbourhoods.

1.3.1.1 Triangular lattices

A 6-curve verifies directly the discrete Jordan's theorem. Figure 1.6 displays an instance of a 6-digital closed curve C, represented with black dots ($\bullet$) linked with bold lines. The points in the interior of C are symbolised by empty squares ($\square$) and the points in the exterior of C are shown as empty circles ($\circ$)

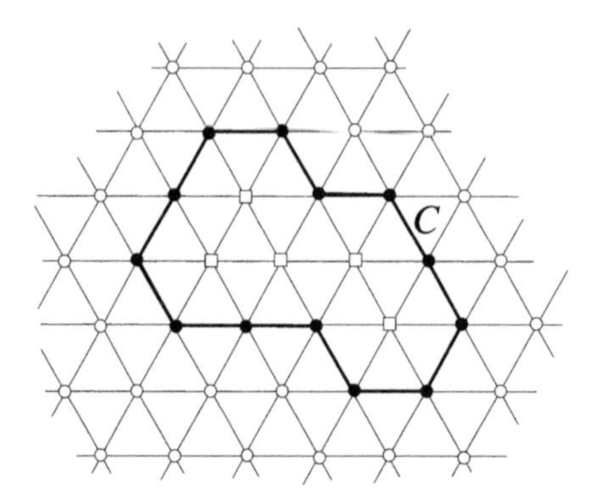

Figure 1.6 A 6-digital closed curve

1.3.1.2 Hexagonal lattices

In contrast to the triangular lattices, the verification of Jordan's theorem on hexagonal lattices is not straightforward. Consider the 3-digital closed curve C shown as bold segments in Figure 1.7. The interior and exterior of this curve cannot be defined using the 3-neighbourhood since there exists no 3-digital arc joining the potential interior points p and q. Therefore, C defines three 3-connected components, one containing p, one containing q and one containing r.

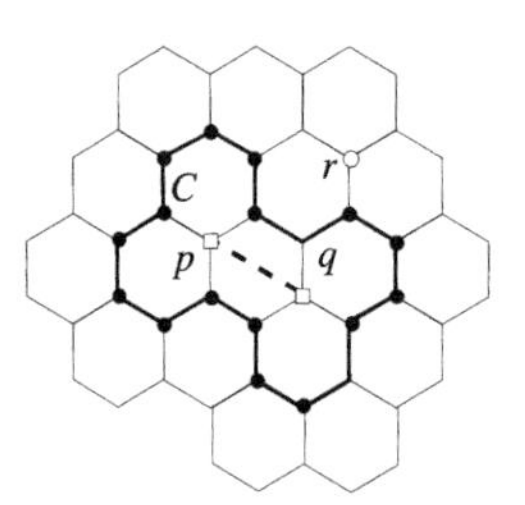

Figure 1.7 A 3-digital closed curve

To overcome this problem, a duality between neighbourhoods on the hexagonal lattice is introduced. If the interior and exterior of a 3-curve are defined as 12-connected components, then Jordan's theorem is verified.

Conversely, the interior and exterior of a 12-curve will be defined as 3-connected components. Figure 1.8 displays an instance where the 12-curve C does not separate the set of points into an interior and an exterior using 12-connectivity. In this example, the potential interior points q and r are both 12-neighbours of p, which is clearly an exterior point. Again, the duality between 3- and 12-neighbourhood is to be used to resolve this problem.

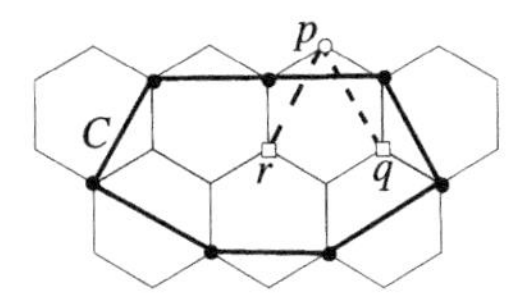

Figure 1.8 A 12-digital closed curve

1.3.1.3 *Square lattices*

Similar cases arise in square lattices. A 4-curve may define more than two 4-connected components. In the example shown in Figure 1.9, three 4-connected components are defined, containing the points p, q and r, respectively.

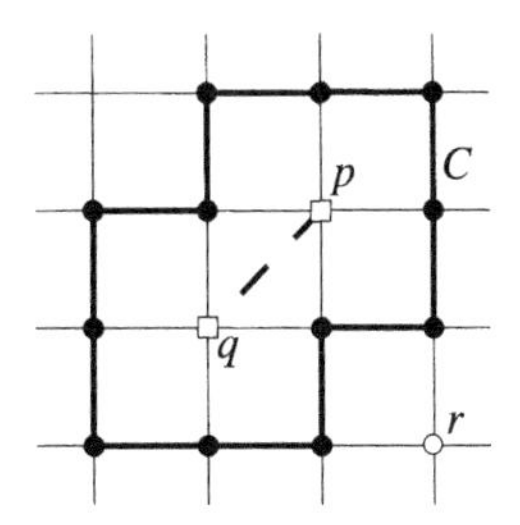

Figure 1.9 A 4-digital closed curve

Such a problem is resolved using 8-connectivity as dual to 4-connectivity. The 4-curve C defines an 8-connected interior component containing the points p and q and the exterior is the 8-connected component containing r. It is clear that there is no 8-arc which connects p or q to any exterior point (i.e. a point in the exterior 8-connected component).

In Figure 1.10, the second type of problem arises. The 8-curve C does not separate the digital plane into two 8-components.

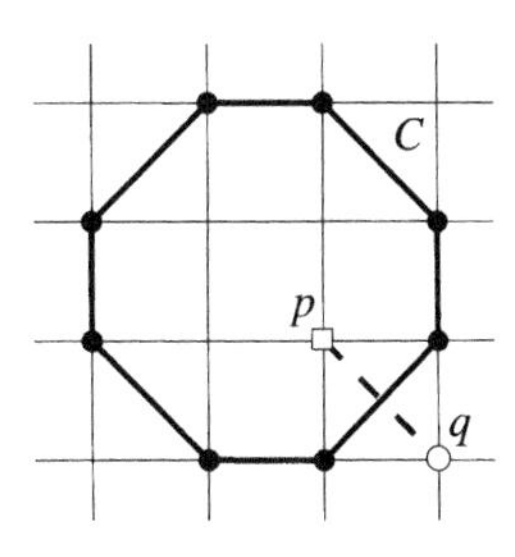

Figure 1.10 A 8-digital closed curve

As a counterexample, there exists an 8-arc joining two potential interior and exterior points p and q, respectively. However, it is clear that an 8-curve will define two 4-connected components as its exterior and interior. Hence, discrete

Jordan's theorem will be satisfied when using 8-connectivity (respectively, 4-connectivity) for the curve and 4-connectivity (respectively, 8-connectivity) for the interior and exterior on the square lattice.

1.3.1.4 Summary

In general, a connectivity relationship cannot be used for both a set and its complement. As shown in the previous sections, a duality between possible (k- and k'-) connectivities and neighbourhoods on the lattices is to be defined. This duality can be summarised as in Table 1.1.

Lattice	Dual connectivities	$(k \leftrightarrow k')$	Dual neighbourhoods
Triangular	6-connectivity $\leftrightarrow$ 6-connectivity	$(6 \leftrightarrow 6)$	$N_6 \leftrightarrow N_6$
Hexagonal	3-connectivity $\leftrightarrow$ 12-connectivity	$(3 \leftrightarrow 12)$	$N_3 \leftrightarrow N_{12}$
Square	4-connectivity $\leftrightarrow$ 8-connectivity	$(4 \leftrightarrow 8)$	$N_4 \leftrightarrow N_8$

Table 1.1 Duality between connectivities and neighbourhoods

Via this duality, neighbourhood relationships are extended to connectivity relationships. Therefore, points can now be grouped in different subsets on which operations are to be performed.

In a topology, sets can be classified either as open or closed sets; the difference is made at the border level. The next section gives definitions and properties of equivalent notions in digital topology.

1.3.2 Border of a digital set

An important subset of points in digital topology is the set of border points which separates a digital set from its complement.

Definition 1.11: Border of a digital set

Given a k-connected set of points P, the complement of P, denoted P^{C}, defines a dual connectivity relationship (denoted k'-connectivity) as given in Table 1.1 (e.g. $k = 8$ and $k' = 4$ for the 8- and 4-connectivities). The border of P is the set of points Γ defined as the k-connected set of points in P that have at least one k'-neighbour (i.e. a neighbour with respect to the k'-connectivity) in P^{C}.

An example for this definition can be given when the set of points represents the pixels in a binary image. A binary image is represented by an array of discrete points labelled with a value (1 or 0) which indicates the black or white colour of the corresponding pixels, respectively. By convention, two basics subsets can be identified.

Definition 1.12: Foreground and background in a binary digital image

(i) The foreground is the set of points F which are labelled with a value equal to 1. By convention, the foreground corresponds to the set of black pixels in a binary image.

(ii) The background is the complement of the set F denoted F^{C}. It is the set of points associated with a zero-value. By convention, the background corresponds to the set of all white pixels in the image.

(iii) Border points are points which form the border of the set according to Definition 1.11. Corresponding pixels in the image are called border pixels. A point (respectively, a pixel) which is not in the border set is referred to as an interior point (respectively, an interior pixel).

Remark 1.13:

The foreground and the background may both contain more than one connected component.

Example: Border of a binary digital image

Consider the digital image shown in Figure 1.11(A). Black pixels (i.e. points of the foreground F) are symbolised as black circles ($\bullet$) and white pixels (i.e. points of the background F^{C}) as white circles ($\circ$). 8-Connectivity is considered in the foreground F and, hence, 4-connectivity is considered in the background (i.e. $k = 8$ and $k' = 4$).

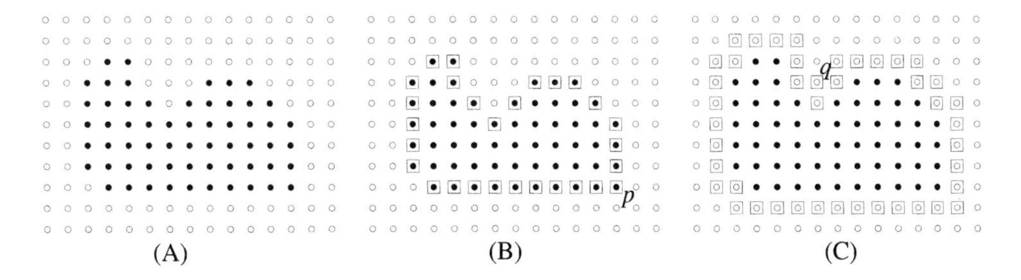

Figure 1.11 Borders of a binary digital image. (A) The representation of the binary image on the square lattice. (B) The foreground is taken as a closed set. (C) The foreground is taken as a open set

Depending on which of the foreground or the background is chosen as an open set, two different border sets are defined. In Figure 1.11(B), the foreground is considered as a closed set. Hence, it contains its border. By definition, the border of the foreground is the set of black pixels Γ that have at least one white pixel among their 4-neighbours. The points in this set are surrounded by a square box in Figure 1.11(B).

Conversely, in Figure 1.11(B), the foreground is considered as an open set. The border therefore belongs to its complement, namely the background. In this case, the border is the set Γ of white pixels that have at least one black pixel among their 8-neighbours. Points in Γ are surrounded by a square box in Figure 1.11(C). This example highlights the fact that the two borders arising from the respective cases are different.

Remark 1.14:

Although the set of border points Γ of a connected component is a connected component with respect to the connectivity of the set it belongs to, it generally does not satisfy the conditions for being a digital closed curve. In the example shown in Figure 1.11(B), Γ, the border of the foreground is 8-connected but the point p in the rightmost bottom corner has three 8-neighbours in Γ. Similarly in Figure 1.11(B), Γ is a 4-connected component. However, the point q has three 4-neighbours in Γ. Therefore, in neither case Γ is a closed curve as defined in Definition 1.8.

Open and closed sets are now well-identified in digital topology. The next step in the construction of a mathematical context for digital image processing is the definition of a metric between points and sets on the lattice. The aim is to define the concept of nearness in a discrete set of points. Such a measurement should be consistent with distance functions defined in the continuous plane. The next section addresses this question by presenting different discrete distance functions which depend on the type of regular lattice and neighbourhood used.

1.4 Discrete distances

By analogy with the continuous space, a discrete distance function should verify the classic metric conditions given in Definition 1.15.

Definition 1.15: Distance

Given a set of points P, a function $d : P \times P \longrightarrow \mathbb{R}^+$ is said to be a distance on P if and only if it satisfies the following conditions.

(i) $d(p,q)$ *is defined and finite for all p and q in P (d is total on P).*

(ii) $d(p,q) = 0$ *if and only if $p = q$ (d is positive definite).*

(iii) $d(p,q) = d(q,p)$, $\forall\,(p,q) \in P \times P$ *(d is symmetric).*

(iv) $d(p,q) + d(q,r) \geq d(p,r)$, $\forall\,(p,q,r) \in P \times P \times P$ *(d satisfies the triangular inequality).*

Distance calculations are based on local distances within neighbourhoods defined in digital topology. Their definitions are related to basic moves on the corresponding lattice as introduced in Definition 1.16.

Definition 1.16: Move and move length

A move on the lattice is the displacement from a point to one of its neighbours. A move length is the value of the local distance between a point and one of its neighbours.

The notion of length for a move can be readily extended to that of a digital arc.

Definition 1.17: Arc length

The length of a digital arc is the sum of the length of the moves that compose it.

A generic definition for a discrete distance is as follows.

Definition 1.18: Discrete distance

Given the lengths for all possible moves in a neighbourhood, the distance between two points p and q is the length of the shortest digital arc (i.e. the digital arc of minimal length) from p to q.

Remark 1.19:

Although the distance between two points is given as a unique value, the digital arc which realises this distance is not necessarily unique.

The fact that such a distance satisfies the metric conditions relies on the definition of move lengths. Historically, a unit value has been attributed to any move length (e.g. see [142]). In this case, the digital arc associated with the distance between p and q is the arc of minimal cardinality joining p and q. Real or integer move lengths have been designed for a discrete distance related to a specific neighbourhood to achieve a close approximation of the Euclidean distance on the plane.

Common definitions of distances are presented here. The case of each of the three possible regular lattices is considered. For each distance, the corresponding discrete disc (Definition 1.20) obtained is also presented. The geometrical properties of such discs constitute an important factor in characterising how closely a discrete distance can approximate the Euclidean distance.

Definition 1.20: Discrete disc

Given a discrete distance d_{D}, a discrete disc of radius $r \geq 0$ centred at point p for this distance is the set of discrete points $\Delta_D(p,r) = \{q \text{ such that } d_{\mathrm{D}}(p,q) \leq r\}$. When no reference to the centre point is necessary, a discrete disc of radius r for the distance d_{D} will also be denoted as $\Delta_D(r)$.

Rounding functions which will be needed throughout this book are formally defined in Definition 1.21.

Definition 1.21: Rounding functions

The rounding, ceiling and floor functions, donated as $\lceil . \rceil$, $\lfloor . \rfloor$ *and* round(.), *respectively, are defined as follows.*

$$\textit{Given } x \in \mathbb{R}, \lceil x \rceil = n \textit{ where } n \in \mathbb{Z} \textit{ is such that } n - 1 < x \leq n.$$

$$\textit{Given } x \in \mathbb{R}, \lfloor x \rfloor = n \textit{ where } n \in \mathbb{Z} \textit{ is such that } n \leq x < n + 1.$$

$$\textit{Given } x \in \mathbb{R}, \mathrm{round}(x) = n \textit{ where } n \in \mathbb{Z} \textit{ is such that } n - \tfrac{1}{2} \leq x < n + \tfrac{1}{2}.$$

1.4.1 Triangular lattices

On the triangular lattice, we define the 6-distance as follows.

Definition 1.22: d_6 distance

The d_6 *distance between p and q is the length of the shortest 6-arc joining p and q when the move lengths in the 6-neighbourhood are all set to the unity.*

Figure 1.12 illustrates a 6-disc of radius $r = 4$ centred at point p. Using the notation introduced in Definition 1.20, this set is denoted $\Delta_6(p, 4)$.

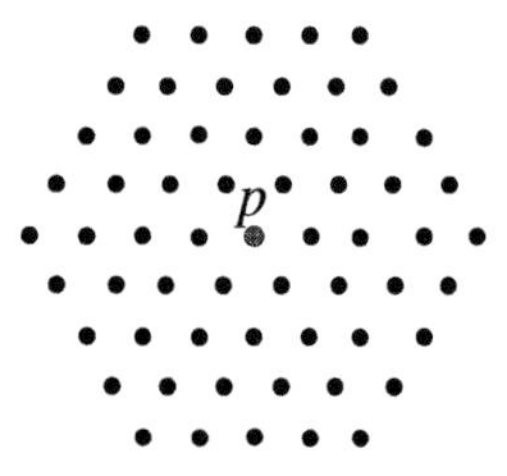

Figure 1.12 A 6-discrete disc of radius 4, centred at p

Remark 1.23:

Note that the convex hull of the disc of radius r for the d_6 *distance is a hexagon of size r (i.e. an expanded version of the 6-neighbourhood). This is due to the fact that the distance propagates evenly in each possible direction, thus preserving the initial shape of the neighbourhood.*

Proposition 1.24:

Given a set of points P on the infinite triangular lattice, d_6 *is a distance on P.*

Proof:
The verification of conditions (i) to (iii) of Definition 1.15 is straightforward. Now, given p, q and r, three points on the triangular lattice, if q is included in the

shortest 6-arc from p to r, then, $d_6(p,r) = d_6(p,q) + d_6(q,r)$. Assume that point q is not included in the shortest 6-arc from p to r. If $d_6(p,q) + d_6(q,r) < d_6(p,r)$ then, there exists a 6-arc which contains q and such that its length is smaller than $d_6(p,r)$. By the definition of $d_6(p,r)$, this is impossible. Therefore, d_6 satisfies the triangular inequality. Hence, Proposition 1.24 holds. □

The 6-neighbourhood allows for the definition of digital objects without further extension (Section 1.3). Moreover, in this neighbourhood, all the moves are equivalent by symmetry. In order to preserve symmetry, it is therefore natural to associate all the moves in the 6-neighbourhood with the same length. Hence, for any such configuration of move lengths, the approximation of the Euclidean distance remains the same. In other words, the best approximation of the Euclidean distance by a discrete distance on the triangular lattice is readily given by d_6. However, extensions to this neighbourhood can be considered for computation purposes (see Section 5.2.3).

1.4.2 Hexagonal lattices

Hexagonal lattices that arise from triangular partitioning are generally not considered for the definition of discrete distances. Figure 1.13(A) clearly shows that the distribution of discrete points induced by hexagonal lattice is not uniform, in contrast with that induced by the square lattice, say (Figure 1.13(B)). For this reason, hexagonal lattices are rarely considered in image processing applications. As a consequence, only few developments have been achieved concerning discrete distances on this type of lattice.

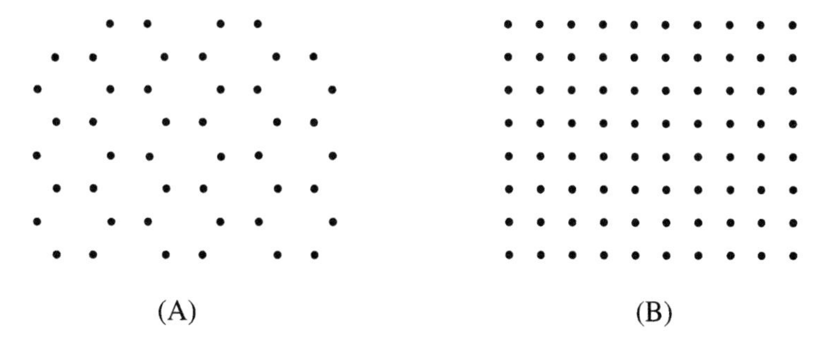

Figure 1.13 Distribution of discrete points on regular lattices. Hexagonal lattice (A) in contrast with square lattice (B)

1.4.3 Square lattices

In the particular case of an infinite square lattice, a point p on the lattice can be uniquely characterised by an integer pair (x_p, y_p) (the coordinates of the point p in the $\mathbb{Z}^2$ plane). Conversely, any integer pair $(x_p, y_p) \in \mathbb{Z}^2$ represents a point on the square lattice. Therefore, there exists a one-to-one mapping from points on the square lattice to $\mathbb{Z}^2$. This property eases the definition of analytical expressions for discrete distances on the square lattice.

Definition 1.25 recalls the analytical expression of the Euclidean distance d_{E} that is used as reference in both continuous and discrete spaces.

Definition 1.25: Euclidean distance

Given two points $p = (x_p, y_p)$ and $q = (x_q, y_q)$ the Euclidean distance value between p and q is given by

$$d_{\mathrm{E}}(p, q) = \sqrt{(x_q - x_p)^2 + (y_q - y_p)^2}$$

It is easy to verify that d_{E} satisfies the conditions to be a distance given in Definition 1.15.

A number of discrete distances have been defined on the square lattice. All move lengths are first set to unity, leading to the d_4 distance (Definition 1.26), d_8 distance (Definition 1.28) and d_{knight} distance (Definition 1.31). Extensions which achieve a closer approximation of the Euclidean distance are also presented in this section.

Definition 1.26: City-block distance

The city-block distance (or Manhattan distance or rectilinear distance) between p and q is the length of the shortest 4-arc joining p and q when the move lengths are all set to unity. The city-block distance between p and q is denoted as $d_4(p, q)$ and is also referred to as the d_4 distance.

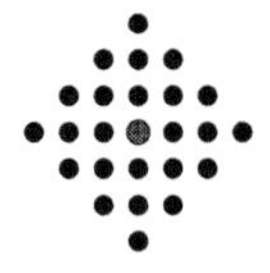

Figure 1.14 4-Disc of radius 3: $\Delta_4(3)$

The location of the points on the square lattice allows for an equivalent definition of the d_4 distance.

Proposition 1.27:

Given two points $p = (x_p, y_p)$ and $q = (x_q, y_q)$ on the square lattice, the minimal cardinality of a 4-arc joining p to q is given by

$$d_4(p, q) = |x_q - x_p| + |y_q - y_p|$$

As a consequence of Proposition 1.27, the 4-neighbourhood of the point p can be characterised by, $\forall\, p = (x_p, y_p) \in \mathbb{Z}^2$,

$$N_4(p) = \{q = (x_q, y_q) \in \mathbb{Z}^2 \text{ such that } |x_q - x_p| + |y_q - y_p| = 1\}$$

More generally, a discrete 4-disc of radius r centred at p (e.g. see Figure 1.14) is characterised by

$$\Delta_4(p, r) = \{q = (x_q, y_q) \in \mathbb{Z}^2 \text{ such that } |x_q - x_p| + |y_q - y_p| \leq r\}$$

A simple extension of the d_4 distance on the 8-neighbourhood leads to the definition of the chessboard distance.

Definition 1.28: Chessboard distance

The chessboard distance (or diamond distance) between p and q is the length of the shortest 8-arc joining p and q when the move lengths are all set to unity. The chessboard distance between p and q is denoted as $d_8(p, q)$ and is also referred to as the d_8 distance.

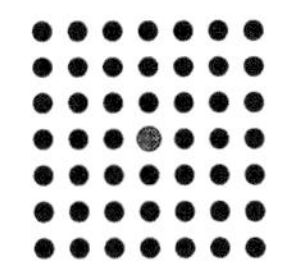

Figure 1.15 8-Disc centred at p and of radius 3: $\Delta_8(p, 3)$

Again, using the coordinates of integer points, d_8 can be given an analytical expression as follows.

Proposition 1.29:

Given two points $p = (x_p, y_p)$ and $q = (x_q, y_q)$ on the square lattice, The minimal cardinality of an 8-arc joining p to q is given by

$$d_8(p, q) = \max(|x_q - x_p|, |y_q - y_p|)$$

From Proposition 1.29, the 8-neighbourhood of the point p can be characterised as follows. $\forall\, p = (x_p, y_p) \in \mathbb{Z}^2$,

$$N_8(p) = \{q = (x_q, y_q) \in \mathbb{Z}^2 \text{ such that } \max(|x_q - x_p|, |y_q - y_p|) = 1\}$$

Therefore, an 8-disc of radius r centred at p is also defined by

$$\Delta_8(p, r) = \{q = (x_q, y_q) \in \mathbb{Z}^2 \text{ such that } \max(|x_q - x_p|, |y_q - y_p|) \leq r\}$$

Since the move lengths that define the d_4 and d_8 distances are all equal to 1, the argument given in the proof of Proposition 1.24 can readily be used to show that both these discrete distance functions satisfy the metric conditions given in Definition 1.15.

Remark 1.30:

Note that there exists a strong similarity between the norms $\|\vec{u}\|_1 = |x_{\vec{u}}| + |y_{\vec{u}}|$, $\|\vec{u}\|_2 = \sqrt{|x_{\vec{u}}|^2 + |y_{\vec{u}}|^2}$ *and* $\|\vec{u}\|_\infty = \sup(|x_{\vec{u}}|, |y_{\vec{u}}|)$ *defined in the continuous space* $\mathbb{R}^2$ *and* d_4, d_E *and* d_8 *on the digital space. Recalling that* $\|\vec{u}\|_1 \leq \|\vec{u}\|_2 \leq \|\vec{u}\|_\infty$ $\forall \vec{u} \in \mathbb{R}^2$, *this property is mapped in the digital space as* $d_4(p,q) \leq d_E(p,q) \leq d_8(p,q)$ $\forall p, q \in \mathbb{Z}^2$.

The *knight*-neighbourhood is studied because of its analogy with the moves of a knight on the chess board. Despite its non-uniform nature, it yields a distance as defined in Definition 1.31.

Definition 1.31: *Knight*-distance

The knight-*distance between p and q is the length of the shortest* knight-*arc joining p and q when the move lengths are all set to unity. The* knight-*distance between p and q is denoted as* $d_{\text{knight}}(p,q)$ *and is also referred to as the* d_{knight} *distance.*

In [30], an alternative analytical characterisation of d_{knight} is given as follows.

Proposition 1.32:

Given two points $p = (x_p, y_p)$ *and* $q = (x_q, y_q)$ *on the square lattice, let* $\delta_{\max} = \max(|x_q - x_p|, |y_q - y_p|)$ *and* $\delta_{\min} = \min(|x_q - x_p|, |y_q - y_p|)$. *Then, the minimal cardinality of a* knight-*arc joining p to q is given by*

$$\begin{aligned} d_{\text{knight}}(p,q) &= \max\left(\left\lceil \frac{\delta_{\max}}{2} \right\rceil, \left\lceil \frac{\delta_{\max} + \delta_{\min}}{3} \right\rceil\right) + (\delta_{\max} + \delta_{\min}) \\ &\quad - \max\left(\left\lceil \frac{\delta_{\max}}{2} \right\rceil, \left\lceil \frac{\delta_{\max} + \delta_{\min}}{3} \right\rceil\right) \bmod 2 \\ &\quad \textit{if } (\delta_{\max}, \delta_{\min}) \neq (1,0) \textit{ and } (\delta_{\max}, \delta_{\min}) \neq (2,2) \\ &= 3 \quad \textit{if } (\delta_{\max}, \delta_{\min}) = (1,0) \\ &= 4 \quad \textit{if } (\delta_{\max}, \delta_{\min}) = (2,2) \end{aligned}$$

The value of $d_{\text{knight}}(p,q)$ given by Proposition 1.32 is not exactly equivalent to the value of $d_{\text{knight}}(p,q)$ derived from Definition 1.31. In [30], d_{knight} is defined as follows. *The distance between any two points is the number of steps taken to trace a shortest path from one point to another.* In other words, $d_{\text{knight}}(p,q)$ is defined as the cardinality of the minimal 4-path which contains all the points in the minimal *knight*-path from p to q. Hence, apart from the special cases of points $(1, 0)$ and $(2, 2)$, Proposition 1.32 results in a value of d_{knight} three times larger than that of Definition 1.31.

It is shown that d_{knight} thus defined satisfies the metric conditions [30]. The concept of *knight*-distance is then extended to that of *super-knight's* distances. All moves contained in the *knight*-neighbourhood can be characterised by the vector $(\delta_x, \delta_y) = (2, 1)$. In super-knight's distance, possible values of the vector (δ_x, δ_y) are investigated for defining moves. Conditions on these values for *super-knight's* distances to satisfy the metric conditions have also been set [31].

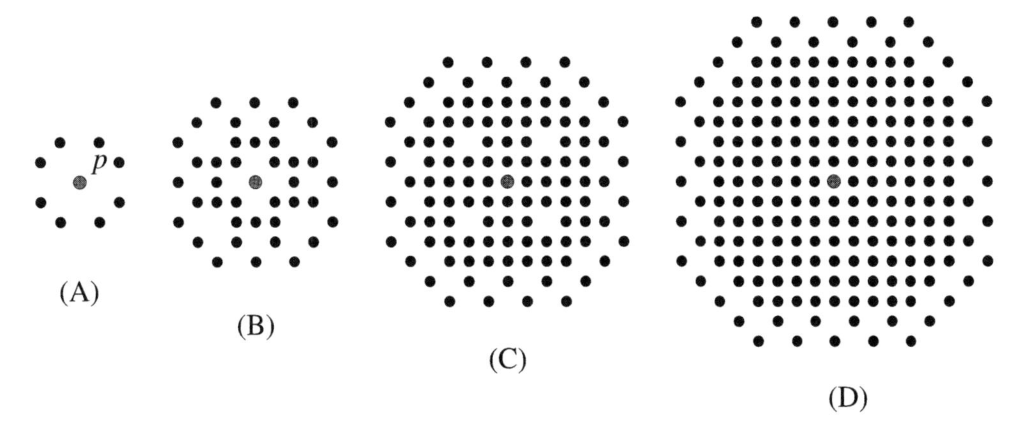

Figure 1.16 *Knight*-discs (following Definition 1.31). (A) $\Delta_{\text{knight}}(p, 1)$. (B) $\Delta_{\text{knight}}(2)$. (C) $\Delta_{\text{knight}}(3)$. (D) $\Delta_{\text{knight}}(4)$

Figure 1.16 illustrates examples of *knight*-discs for different values of the radius. Unlike other discrete discs, *knight*-discs exhibit a non-regular evolution of their structure. Moreover, because of the "incomplete" nature of the *knight*-neighbourhood, *knight*-discs have some points missing close to their borders. For a detailed study of *knight*- and *super knight's*- distances, the reader is referred to references [30, 31].

Combining the 8-neighbourhood with the *knight*-neighbourhood, thus forming the 16-neighbourhood with unit move lengths, does not yield a distance (see Remark 1.39). These neighbourhoods are therefore not detailed further at this stage.

The hexagonal disc induced by d_6 on the triangular lattice is the best approximation of the Euclidean disc obtained so far. On the other hand, the square lattice offers the advantage of analytical distance calculations. In order to combine both advantages, the hexagonal distance is defined on the square lattice [142].

Definition 1.33: Hexagonal distance

Given two points $p = (x_p, y_p)$ and $q = (x_q, y_q)$ on the square lattice, the hexagonal *distance* $d_{\text{hex}}(p,q)$ *from* p *to* q *is given by*

$$
\begin{aligned}
d_{\text{hex}}(p,q) = \max \Big(& |x_q - x_p|, \\
& \frac{1}{2}(|x_q - x_p| + (x_p - x_q)) - \left(\left[\frac{x_p}{2}\right] - \left[\frac{x_q}{2}\right]\right) + y_q - y_p, \\
& \frac{1}{2}(|x_q - x_p| + (x_p - x_q)) - \left(\left[\frac{x_p}{2}\right] - \left[\frac{x_q}{2}\right]\right) + y_p - y_q \Big)
\end{aligned}
$$

where $[x]$ *denotes the integer part of* x.

The hexagonal distance corresponds to d_6 where points on the triangular lattice are mapped onto the square lattice as shown in Figure 1.17.

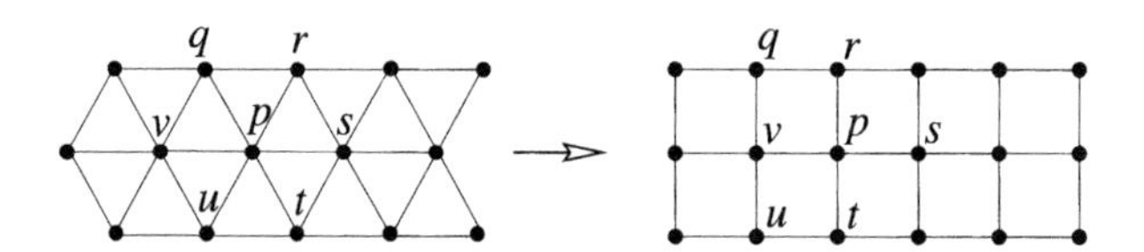

Figure 1.17 Mapping from the triangular to the square lattice

The d_{hex} distance can be characterised by the *hex*-disc shown in Figure 1.18(A). Points are shown as figures that represent the values of the relevant distance from the centre point. For comparison, the equivalent 6-disc ($\Delta_6(p, 4)$) is shown in Figure 1.18(B). Figure 1.18 also highlights the fact that the hexagonal distance is defined as a path which combines 4- and 8-moves. This combination is the exact mapping of the 6-path on the triangular lattice.

```
        4 4 4 4 4                    4 4 4 4 4
      4 3 3 3 3 4                  4 3 3 3 3 4
      4 3 2 2 2 3 4              4 3 2 2 2 3 4
  4 3 2 1 1 2 3 4              4 3 2 1 1 2 3 4
  4 3 2 1 0 1 2 3 4          4 3 2 1 0 1 2 3 4
  4 3 2 1 1 2 3 4              4 3 2 1 1 2 3 4
      4 3 2 2 2 3 4              4 3 2 2 2 3 4
      4 3 3 3 3 4                  4 3 3 3 3 4
        4 4 4 4 4                    4 4 4 4 4

           (A)                          (B)
```

Figure 1.18 Hexagonal discs on the square lattice. (A) *Hex*-disc of radius 4 ($\Delta_{\text{hex}}(4)$). (B) The equivalent 6-disc, $\Delta_6(4)$

Similarly, the octagonal distance d_{oct} is defined as a combination of the d_4 and d_8 distances [142]. 4-Moves and 8-moves are alternatively allowed on the square lattice. The analytical expression for d_{oct} is given in Definition 1.34.

Definition 1.34: Octagonal distance

Given two points $p = (x_p, y_p)$ and $q = (x_q, y_q)$ on the square lattice, the octagonal distance $d_{\text{oct}}(p, q)$ from p to q is given by

$$d_{\text{oct}}(p,q) = \max\left(d_8(p,q), \left[\frac{2}{3}(|x_q - x_p| + |y_q - y_p| + 1)\right]\right)$$

where $[x]$ denotes the integer part of x.

For example, using this definition, a resulting *oct*-disc of radius 4 is shown in Figure 1.19.

		4	4	4	4	4		
	4	4	3	3	3	4	4	
4	4	3	2	2	2	3	4	4
4	3	2	2	1	2	2	3	4
4	3	2	1	0	1	2	3	4
4	3	2	2	1	2	2	3	4
4	4	3	2	2	2	3	4	4
	4	4	3	3	3	4	4	
		4	4	4	4	4		

Figure 1.19 *Oct*-disc of radius 4, $\Delta_{\text{oct}}(4)$

Although *hex*- and *oct*-discs represent an improvement in the approximation of Euclidean discs by discrete discs on the square lattice, their definitions are not simple. As a consequence, the study of properties for such distances is not immediate. Moreover, no possible extension is clearly suggested by the definitions of the hexagonal and octagonal distances.

With the aim of improving simplicity and accuracy in the approximation of Euclidean distance on the square lattice, chamfer distances have been introduced as a generalisation of all previous definitions. In chamfer discrete distances, moves are given different lengths depending on some criteria. Chamfer distances have been intensively studied for developing image processing operators. The generic definition of a chamfer distance is given as follows.

Definition 1.35: Chamfer distance

Given a neighbourhood and associated move lengths, the chamfer distance between p and q relative to this neighbourhood is the length of the shortest digital arc from p to q.

A chamfer distance is relative to a neighbourhood associated with move lengths. Chamfer distances build on the neighbourhoods already presented in Section 1.2. A general scheme for constructing neighbourhoods is then derived and the general definition of the chamfer distance is given.

Starting with the 4-neighbourhood, the length of a 4-move is denoted a. In this respect, a 4-move is also called an a-move. Clearly, in the 4-neighbourhood, all moves are equivalent by symmetry or rotation. In this case, the only possible definition of a discrete distance that is geometrically consistent is that of the d_4 distance, where $a = 1$.

A simple extension of the 4-neighbourhood leads to the 8-neighbourhood. Diagonal moves are added to the horizontal and vertical moves. The length of such diagonal moves is denoted b. In this respect, diagonal moves are called b-moves and the chamfer distance obtained in the 8-neighbourhood is denoted $d_{a,b}$. Given any positive value for a (i.e. the length for all 4-moves), in order to preserve a geometrical consistency within the 8-neighbourhood, the diagonal moves should be associated with a length b larger than a. In this context, the most natural value is $b = a\sqrt{2}$, since it allows for an exact value of the chamfer distance along the diagonal lines from a given point. However, for the sake of simplicity of computation and storage, it is also important to preserve integer arithmetic for distance calculations. In this respect, integer values for a and b have been derived (e.g. see [12,13,57,97]). The most commonly used set of such values is $a = 3$ and $b = 4$ [12,13].

Referring to Section 1.2, a further extension defines the 16-neighbourhood. The *knight*-move is introduced and its length is denoted c (thus defining a c-move). The chamfer distance obtained in the 16-neighbourhood is denoted $d_{a,b,c}$. Assuming that $a = 1$, $b = \sqrt{2}$, the value $c = \sqrt{5}$ allows for an exact chamfer distance value along the lines that support the c-moves. For preserving integer calculations of chamfer distances, the lengths of the moves included in the 16-neighbourhood (1, $\sqrt{2}$, $\sqrt{5}$) are commonly approximated by using the set of integer values ($a = 5$, $b = 7$, $c = 11$) [12,13].

The fact that a chamfer distance satisfies the metric conditions given in Definition 1.15 depends on the values of the move lengths. Hence, restrictions on these values for chamfer distances to satisfy the metric conditions have been set.

Proposition 1.36:

The conditions a and b for $d_{a,b}$ to be a discrete distance are $0 < a \leq b \leq 2a$.

The typical values of $a = 3$ and $b = 4$ satisfy these conditions and therefore $d_{3,4}$ is a distance in the 8-neighbourhood. In this case, the value of the diagonal move length $\sqrt{2}$ is approximated by $\frac{4}{3}$.

Remark 1.37:

Note that the values $a = b = 1$ used for the definition of d_8 satisfy the conditions given in Proposition 1.36. Therefore, d_4 and d_8 can be seen as particular cases of chamber distances in the 4- and 8-neighbourhoods, respectively.

Similar conditions can be expressed in the 16-neighbourhood for $d_{a,b,c}$ to be a discrete distance.

Proposition 1.38:

The values of a, b and c should satisfy the following conditions for $d_{a,b,c}$ to be a distance on the 16-neighbourhood.

$$0 < a \leq b \leq 2a \leq c \quad \text{and} \quad c \leq a + b \quad \text{and} \quad 3b \leq 2c$$

Again, the typical values for the move lengths $a = 5$, $b = 7$ and $c = 11$ satisfy the above conditions. Therefore, $d_{5,7,11}$ is a distance. In this case, the diagonal move length $\sqrt{2}$ is approximated by $\frac{b}{a} = \frac{7}{5}$ and the *knight*-move length of $\sqrt{5}$ is approximated by $\frac{c}{a} = \frac{11}{5}$.

Remark 1.39:

The values $a = b = c = 1$ do not satisfy the conditions given in Proposition 1.38. Therefore, as mentioned earlier, an extension of d_8 in the 16-neighbourhood by setting all move lengths to unity is not possible.

Chamfer discs are presented in Figure 1.20. Typically, the convex hull of a chamfer disc in the 8-neighbourhood is an octagon that approximates the Euclidean circle depending on the values of a and b. More generally, a chamfer disc is a polygon with as many sides as there are different moves in the neighbourhood on which the chamfer distance is defined.

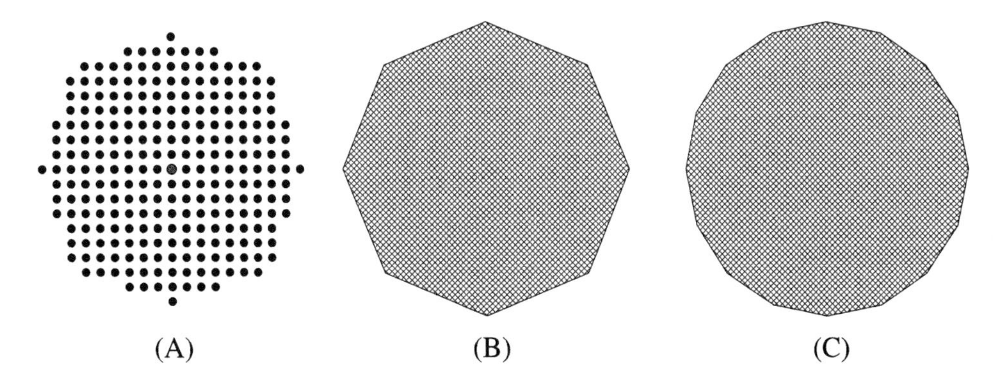

Figure 1.20 Chamfer discs. (A) $\Delta_{3,4}(27)$. (B) $\Delta_{a,b}$. (C) $\Delta_{a,b,c}$

Chamfer distances allow for an easy extension of the previous instances. This extension relies on the creation of size-increasing neighbourhoods on the square lattice. Moves are ordered with the slope of the line that they follow and successively added to the neighbourhood. The Farey sequences that give the order of such slopes forms the basis of the neighbourhood extension.

Because of the symmetry induced on the square lattice, only moves included in the first octant (i.e. slopes included between 0 and 1) are detailed here. Given a set of moves in the first octant, the complete neighbourhood is readily obtained by symmetry or rotation. A horizontal move follows a line of slope $\delta_y/\delta_x = 0/1 = 0$ in the plane. Similarly, a diagonal move follows a line of slope $\delta_y/\delta_x = 1$.

Now, consider a *knight*-move in this context, it follows a line of slope $\delta_y/\delta_x = 1/2$. The sequence of directions for all possible moves in the 16-neighbourhood (i.e. combining the 8- and the *knight*-neighbourhoods) is therefore $\delta_y/\delta_x \in \left\{0, \frac{1}{2}, 1\right\}$. This set is called Farey sequence of order 2. Farey sequences can be used to insert successive rational numbers in this set as slopes of the successive moves considered. Formally, Farey sequences are defined as follows.

Definition 1.40: Farey sequence

The Farey sequence of order $n > 0$ is the set F_n of rational numbers

$$F_n = \left\{\frac{i}{j} \textit{ such that } 0 < j \leq n \; ; \; 0 \leq i \leq j \; ; \; \gcd(i,j) = 1\right\}$$

where $\gcd(i,j)$*, the greatest common divisor of i and j is defined by*

$$\gcd(i,j) = \max_{\mathbb{Z}}\{l \;/\; \exists\, i' \in \mathbb{Z}\,,\; j' \in \mathbb{Z} \textit{ such that } i = i'.l \textit{ and } j = j'.l\}$$

Example: Farey sequence of order 6

The Farey sequence of order 6 is the set of rational numbers

$$F_6 = \left\{0, \frac{1}{6}, \frac{1}{5}, \frac{1}{4}, \frac{1}{3}, \frac{2}{5}, \frac{1}{2}, \frac{3}{5}, \frac{2}{3}, \frac{3}{4}, \frac{4}{5}, \frac{5}{6}, 1\right\}$$

Remark 1.41:

The following property of Farey sequence can be proved using number theory and is easily verified using the example above. Given a Farey sequence F_n, for any two consecutive terms $\frac{i}{j} \in F_n$ and $\frac{i'}{j'} \in F_n$ such that $\frac{i}{j} < \frac{i'}{j'}$, $i'j - ij' = 1$.

Extensions of the 16-neighbourhood can therefore be defined using Farey sequences. Figure 1.21 illustrates the neighbourhoods associated with the Farey sequence of order 3 (i.e. $\delta_y/\delta_x \in F_3 = \left\{0, \frac{1}{3}, \frac{1}{2}, \frac{2}{3}, 1\right\}$) and 4 (i.e. $\delta_y/\delta_x \in F_4 = \left\{0, \frac{1}{4}, \frac{1}{3}, \frac{1}{2}, \frac{2}{3}, \frac{3}{4}, 1\right\}$). The minimal moves following the lines of slopes $\delta_y/\delta_x = \frac{1}{3}$ and $\delta_y/\delta_x = \frac{2}{3}$ (and their symmetric counterparts) are added to the 16-neighbourhood from the Farey sequence F_3 (see Figure 1.21(A)). Then, moves following the slopes $\delta_y/\delta_x = \frac{1}{4}$ and $\delta_y/\delta_x = \frac{3}{4}$ (and their symmetric counterparts) are added from the Farey sequence F_4 (see Figure 1.21(B)). Using such an extension scheme, moves can be defined on the square lattice. Following the notation for move length, the length of the moves added by the Farey sequence F_3 is denoted d (slope $\delta_y/\delta_x = \frac{1}{3}$), e (slope $\delta_y/\delta_x = \frac{2}{3}$) and so on.

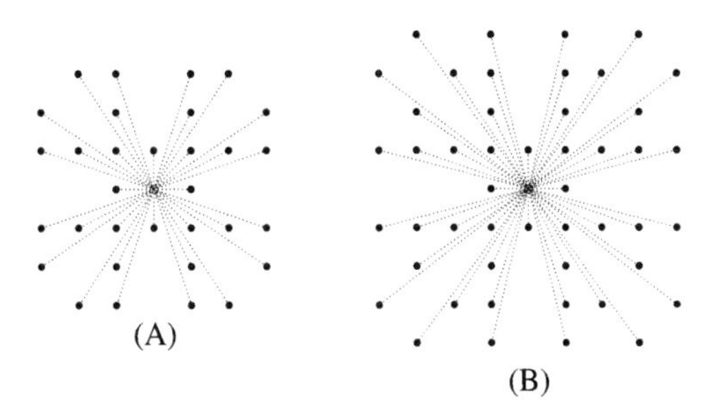

Figure 1.21 Neighbourhoods corresponding to Farey sequences. (A) Order 3. (B) Order 4

Based on these moves, a general definition of chamfer distances can be given as follows.

Definition 1.42: Chamfer distance

Given a neighbourhood related to a Farey sequence F_n and the associated move lengths (a, b, c,...), the chamfer distance between p and q relative to this neighbourhood is the length of the shortest digital arc from p to q.

General conditions on move lengths can be defined for chamfer distances to satisfy the metric properties [162].

Proposition 1.43:

Given a neighbourhood on the square lattice induced by $\{\delta_y^i/\delta_x^i \in F_n\}$ and associated move lengths l_i (i.e. $l_i = a, b, c, \ldots$), then, the resulting chamfer distance satisfies the metric conditions given in Definition 1.15 if and only if, for all i

$$\frac{\delta_x^{i+1}}{\delta_x^i} l_i \leq l_{i+1} \leq \frac{\delta_x^{i+1} + \delta_y^{i+1}}{\delta_x^i + \delta_y^i} l_i \quad \textit{and} \quad \theta_i^{i+2} l_{i+1} \leq \theta_{i+1}^{i+2} l_i \leq \theta_i^{i+1} l_{i+2}$$

where, $\theta_m^l = \delta_x^m \delta_y^l - \delta_x^l \delta_y^m$

It is easy to verify that all conditions on a, b and c given in Propositions 1.36 and 1.38 are particular instances of the general form given in Proposition 1.43 above.

1.4.4 Summary

Discrete distances have been presented for all possible regular lattices. They represent approximations of the Euclidean distance function in the continuous plane. The use of a particular discrete distance depends on the application considered.

The hexagonal lattice is derived from a triangular partition of the plane. Such a partition is not, in general, a realistic representation of a physical captor device. Moreover, as noted in Section 1.4.2, the hexagonal lattice exhibits a non-uniform distribution of discrete points in the plane (see Figure 1.13). Consequently, the hexagonal lattice is rarely used in digital image processing applications.

The triangular lattice offers a simple and complete neighbourhood (i.e. the 6-neighbourhood). Its particular structure allows for the reduction of inconsistencies induced by the acquisition process (e.g. aliasing). Moreover, the hexagonal partitioning is the best approximation of retinal cells and this structure is often used in acquisition or display devices (e.g. video camera or television). Nevertheless, the square lattice allows for a simple analytical calculation of discrete distance values. The square lattice is, therefore, generally preferred to any other lattice. The triangular lattice can be implicitly used through a mapping onto the square lattice (e.g. see Figure 1.17). Chamfer distances defined on the square lattice can be formally defined and are the most studied distance functions in digital image processing.

In summary, hexagonal partitions (and hence triangular lattices) are generally preferred for physical reasons but most theoretical work uses the square lattice (and the square partition). In the remaining part of this book, we will concentrate on square lattices and particularly on the dual 4- and 8-neighbourhoods (i.e. using d_4, d_8 and $d_{a,b}$). When possible, extensions to other neighbourhoods or lattices will be presented and developed.

1.5 Compatibility with continuous distances

In Sections 1.1 to 1.3, a model for digital topology has been developed. Based on this, continuous concepts such as continuity and distance have been mapped onto regular lattices as the concepts connectivity and discrete distance. These notions set the theoretical context for the definition of operators in the discrete space.

Discrete distances presented in Section 1.4 represent an approximation of the Euclidean distance in the continuous plane. The natural embedding of $\mathbb{Z}^2$ into $\mathbb{R}^2$ would allow for the use of the Euclidean distance in digital topology. However, the use of discrete distances is generally recommended for the following reasons.

Discrete distances can be defined using local integer coefficients and distance calculations can therefore be computed using integer arithmetic. This implies an important reduction of the computational effort induced by image processing operators. Similarly, the use of integer arithmetic eases the storage of distance values. Finally, by definition, discrete distance can be computed by local operations. In contrast, such a property is not valid when using Euclidean distance. This property will prove fundamental when developing algorithms for image processing.

With this in mind, it is of interest to quantify and optimise the quality of the approximation when using discrete distances. Two criteria can be considered in the study of such an approximation. Firstly, it is important to know the approximation error made when calculating a distance between any two points. This problem is addressed in Section 1.5.1. Secondly, as result of this approximation, operations such as rotations may induce geometrical inconsistencies. In Section 1.5.2, a general approach for the study of such problems is summarised.

1.5.1 Approximation errors

Discrete discs can be represented by polygons with a number of sides varying with the size of the neighbourhood on which the distance in defined (e.g. see Figure 1.20). For any set of (real or integer) move lengths (e.g. a, b, c), an error still remains in the approximation of Euclidean discs by discrete discs. Increasing the size of the neighbourhood partially resolves this problem but also increases the computational effort required for distance computations.

The relative approximation error between Euclidean and discrete distances depends on the value of the move lengths. Optimisation schemes have been proposed to reduce this error. They are mostly based on the definition of a quality criterion and the calculation of optimal move lengths which match this criterion for any given neighbourhood size. A classic scheme aims for the reduction of the approximation error along a given line or circle. We sketch the calculation of the relative error resulting from $d_{a,b}$ and $d_{a,b,c}$ along a vertical line.

Definition 1.44: Relative error

The relative error between the values of a given discrete distance d_{D} *and the Euclidean distance* d_{E} *between two points* O *and* p *is calculated as*

$$E_{\mathrm{D}}(O,p) = \frac{(1/\varepsilon)d_{\mathrm{D}}(O,p) - d_{\mathrm{E}}(O,p)}{d_{\mathrm{E}}(O,p)} = \frac{1}{\varepsilon}\left(\frac{d_{\mathrm{D}}(O,p)}{d_{\mathrm{E}}(O,p)}\right) - 1$$

Parameter $\varepsilon > 0$ is called the scale factor. It is used to maintain consistency between radii of discrete and Euclidean discs. When using chamfer distances, a typical value is $\varepsilon = a$. In this case, the discrete distance value along the horizontal and vertical directions from a given point is exact.

Let $d_D = d_{a,b}$. Without loss of generality, the origin is set in O (i.e. $(x_O, y_O) = (0,0)$). Considering the first octant (i.e. between the lines $y = 0$ and $y = x$), with the aid of Figure 1.22, it is easily seen that

$$d_{a,b}(O,p) = (x_p - y_p)a + y_p b$$

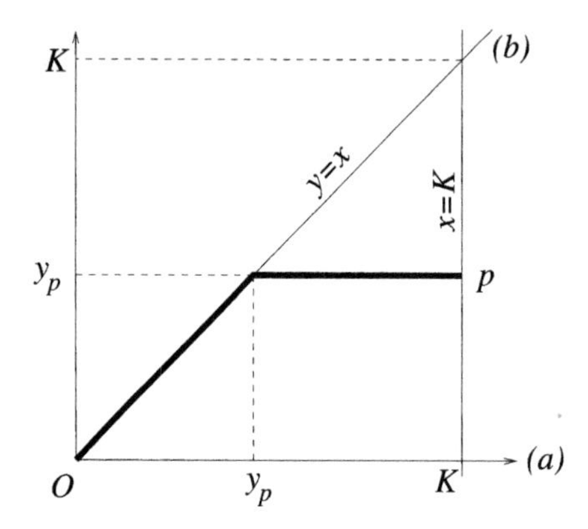

Figure 1.22 Calculation of $d_{a,b}$ in the first octant

The error $E_{a,b}$ is measured along the line $(x = K)$ with $K > 0$ (i.e. for all p such that $x_p = K$ and $0 \leq y_p \leq K$). Therefore, using Definition 1.44, for all p such that $x_p = K$ and $0 \leq y_p \leq K$, the value of the relative error at point p is given by

$$E_{a,b}(O,p) = \frac{(K - y_p)a + y_p b}{\varepsilon\sqrt{K^2 + y_p^2}} - 1$$

Typically, the graph of $E_{a,b}$ for $y_p \in [0, K]$ is as shown in Figure 1.23. On the upper horizontal axis, the names of the move directions (i.e. a, b) met by the line $(x = K)$ are indicated between brackets.

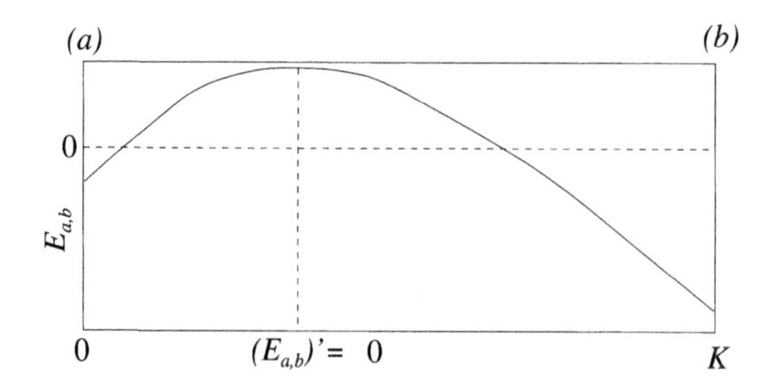

Figure 1.23 The graph of $E_{a,b}$ in the first octant

Since $E_{a,b}$ is a convex function on $[0, K]$, its local extreme can be obtained at p such that $\frac{\partial E_{a,b}}{\partial y}(O,p) = 0$. Now, for all p such that $x_p = K$ and $0 \leq y_p \leq K$,

$$\frac{\partial E_{a,b}}{\partial y}(O,p) = \frac{1}{\varepsilon\sqrt{K^2 + y_p^2}}\left((b-a) - \frac{((K-y_p)a + y_p b)y_p}{K^2 + y_p^2}\right)$$

and $\frac{\partial E_{a,b}}{\partial y}(O,p) = 0$ if $p = (K, \frac{b-a}{a}K)$. In that case, $E_{a,b}(O,p) = \frac{\sqrt{a^2+(b-a)^2}}{\varepsilon} - 1$.

The maximum relative error defined as $E_{a,b}^{\max} = \max\{|E_{a,b}(O,p)|\ ;\ p \,/\, x_p = K\ ;\ 0 \leq y_p \leq K\}$ is then either reached at the local extreme or at a bound of the interval $y_p \in [0, K]$ (see Figure 1.23). Now, $E_{a,b}(O,p) = \frac{a}{\varepsilon} - 1$ if $p = (K, 0)$ and $E_{a,b}(O,p) = \frac{b}{\varepsilon\sqrt{2}} - 1$ if $p = (K, K)$.

Hence, $E_{a,b}^{\max}(O,p) = \max\left(\left|\frac{a}{\varepsilon} - 1\right|, \left|\frac{\sqrt{a^2+(b-a)^2}}{\varepsilon} - 1\right|, \left|\frac{b}{\varepsilon\sqrt{2}} - 1\right|\right)$.

Remark 1.45:

Although the error is calculated along a line rather than a circle, its value does not depend on the line (i.e. $E_{a,b}^{\max}$ does not depend on K). Hence, the value of $E_{\max}^{a,b}$ is valid throughout the discrete plane $\mathbb{Z}^2$.

For $a = 3$, $b = 4$ and $\varepsilon = a = 3$ the maximal relative error is

$$E_{3,4}^{\max} = \left|\frac{4}{3\sqrt{2}} - 1\right| < 5.73\%$$

The study of $d_{a,b,c}$ in the first octant involves two cases, each similar to that of $d_{a,b}$. The calculation of the relative error $E_{a,b,c}(o,p)$ is again achieved for all points p such that $x_p = K > 0$. Figure 1.24 highlights the fact that, depending on whether $0 \leq y_p \leq \frac{K}{2}$ or $\frac{K}{2} \leq y_p \leq K$, the value of $d_{a,b,c}$ and hence $E_{a,b,c}(O,p)$ can be expressed differently.

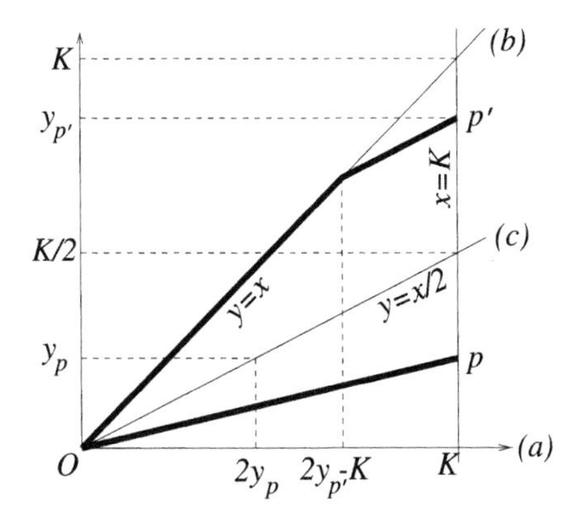

Figure 1.24 Calculation of $d_{a,b,c}$ in the first octant

If $0 \leq y_p \leq \frac{K}{2}$, $d_{a,b,c}(O,p) = (K - 2y_p)a + y_p c$. Then, for all p such that $x_p = K$ and $0 \leq y_p \leq \frac{K}{2}$,

$$E_{a,b,c}(O,p) = \frac{(K - 2y_p)a + y_p c}{\varepsilon\sqrt{K^2 + y_p^2}} - 1$$

and

$$\frac{\partial E_{a,b,c}}{\partial y}(O,p) = \frac{1}{\varepsilon\sqrt{K^2 + y_p^2}}\left((c - 2a) - \frac{((K - 2y_p)a + y_p c)y_p}{K^2 + y_p^2}\right)$$

Similarly, if $\frac{K}{2} \leq y_p \leq K$, $d_{a,b,c}(O,p) = (2y_p - K)b + (K - y_p)c$. Then, for all p such that $x_p = K$ and $\frac{K}{2} \leq y_p \leq K$,

$$E_{a,b,c}(O,p) = \frac{(2y_p - K)b + (K - y_p)c}{\varepsilon\sqrt{K^2 + y_p^2}} - 1$$

and

$$\frac{\partial E_{a,b,c}}{\partial y}(O,p) = \frac{1}{\varepsilon\sqrt{K^2 + y_p^2}}\left((2b - c) - \frac{((2y_p - K)b + (K - y_p)c)y_p}{K^2 + y_p^2}\right)$$

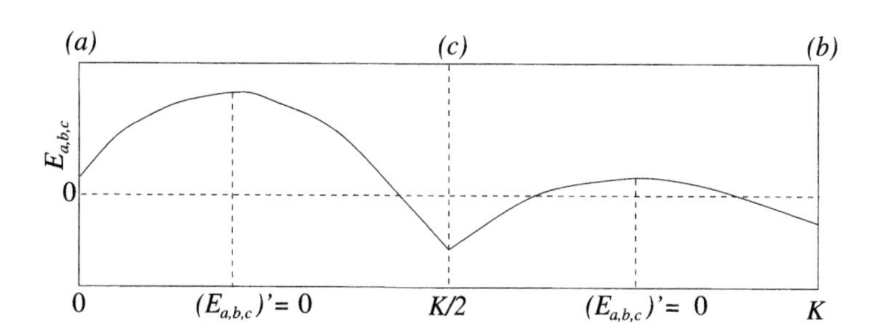

Figure 1.25 The graph of $E_{a,b,c}$ in the first octant

Figure 1.25 illustrates a typical graph for $E_{a,b,c}$ and suggests that the maximum relative error $E_{a,b,c}^{\max} = \max(|E_{a,b,c}(O,p)|\ ;\ p\ /\ x_p = K\ ;\ 0 \leq y_p \leq K)$ is obtained for one of the five following values of y_p:

- $y_p = 0$: $p = (K, 0) \Rightarrow E_{a,b,c}(O,p) = \frac{a}{\varepsilon} - 1$.
- $0 < y_p < \frac{K}{2}$ and $\frac{\partial E_{a,b,c}}{\partial y}(O,p) = 0$: $p = (K, \frac{c-2a}{a}K) \Rightarrow E_{a,b,c}(O,p) = \frac{\sqrt{a^2+(c-2a)^2}}{\varepsilon} - 1$.

- $y_p = \frac{K}{2}$: $p = (K, \frac{K}{2}) \Rightarrow E_{a,b,c}(O,p) = \frac{c}{\varepsilon\sqrt{5}} - 1$.
- $\frac{K}{2} < y_p < K$ and $\frac{\partial E_{a,b,c}}{\partial y}(O,p) = 0$: $p = (K, \frac{2b-c}{c-b}K) \Rightarrow E_{a,b,c}(O,p) = \frac{\sqrt{(c-b)^2+(2b-c)^2}}{\varepsilon} - 1$.
- $y_p = K$: $p = (K,K) \Rightarrow E_{a,b,c}(O,p) = \frac{b}{\varepsilon\sqrt{2}} - 1$.

Therefore,

$$E^{\max}_{a,b,c} = \max\left(\left|\frac{a}{\varepsilon} - 1\right|, \left|\frac{\sqrt{a^2+(c-2a)^2}}{\varepsilon} - 1\right|, \left|\frac{c}{\varepsilon\sqrt{5}} - 1\right|, \left|\frac{\sqrt{(c-b)^2+(2b-c)^2}}{\varepsilon} - 1\right|, \left|\frac{b}{\varepsilon\sqrt{2}} - 1\right|\right)$$

For $(a,b,c) = (5,7,11)$ and $\varepsilon = a = 5$,

$$E^{\max}_{5,7,11} = \left|\frac{\sqrt{26}}{5} - 1\right| < 1.99\%$$

Based on this error calculation scheme, Table 1.2 summarises the improvement achieved in the approximation of Euclidean distance obtained by using chamfer distances.

Distance	Neighbour-hood	Move lengths and scale factor	Maximum relative error
d_4	N_4	$a = 1,\ \varepsilon = 1$	41.43%
d_8	N_8	$a = 1,\ b = 1,\ \varepsilon = 1$	29.29%
$d_{2,3}$	N_8	$a = 2,\ b = 3,\ \varepsilon = 2$	11.80%
$d_{3,4}$	N_8	$a = 3,\ b = 4,\ \varepsilon = 3$	5.73%
$d_{5,7,11}$	N_{16}	$a = 5,\ b = 7,\ c = 11,\ \varepsilon = 5$	1.99%
$d_{72,102,161}$	N_{16}	$a = 72,\ b = 102,\ c = 161,\ \varepsilon = 73^{(*)}$	1.33%

(*)Optimal move lengths and scale factor for N_{16} proposed in [162].

Table 1.2 Maximum relative errors for some instances of move lengths

From the previous two examples of $E_{a,b}$ and $E_{a,b,c}$, a general scheme for the calculation of the maximal relative error $E^{\max}_{\mathrm{D}}$ induced by a discrete distance d_{D} in a given neighbourhood can readily be defined. The first octant is split in a number of sections, in each of which only two different moves interact. Analytical expressions for the error in each section can then be defined and the value of $E^{\max}_{\mathrm{D}}$ is readily defined as the global maximum. The optimisation of move lengths within a given neighbourhood can then be achieved by iterating error calculations with different value of the move lengths. Move lengths which are said to be optimal for the approximation of the Euclidean distance have been proposed [4, 13, 15, 162, 166, 168]. A trade-off is then to be defined between the quality of the approximation and the computational effort induced by the range of the move lengths and the size of the neighbourhood considered.

The next section introduces another criterion which has to be considered in the development of discrete operators.

1.5.2 Geometrical inconsistencies

Inconsistencies that arise during geometric transformations can be summarised in the following example. Consider the set of discrete points highlighted (•) in Figure 1.26.

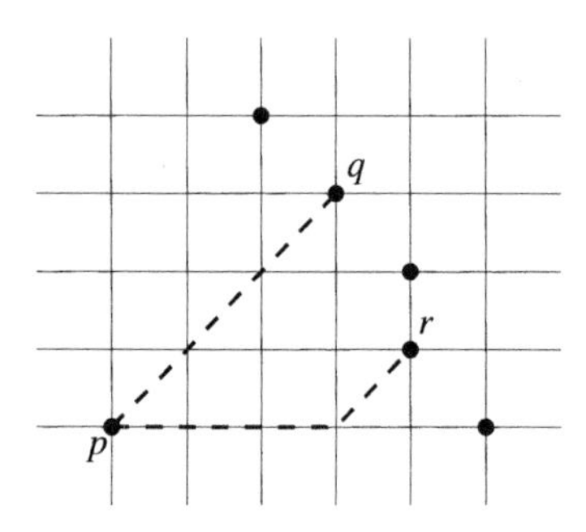

Figure 1.26 Geometrical inconsistencies in the discrete distance calculations

Let the discrete distance used be the chamfer distance $d_{3,4}$ (i.e. $a = 3$ and $b = 4$). When using the discrete distance $d_{a,b}$, p is closer to q than any other of the highlighted points (e.g. point r) since $d_{3,4}(p, q) = 12$ and $d_{3,4}(p, r) = 13$. However, it is easily seen that $d_{\mathrm{E}}(p, q) = \sqrt{18}$ and $d_{\mathrm{E}}(p, r) = \sqrt{17}$. Therefore, p, q and r are such that $d_{3,4}(p, q) < d_{3,4}(p, r)$ and $d_{\mathrm{E}}(p, r) > d_{\mathrm{E}}(p, q)$. Therefore, not only are the discrete distance values approximations of the Euclidean distance, but the ordering induced by the discrete distance does not match that of the Euclidean distance. As a consequence, geometrical inconsistencies arise in the definition of discrete operators.

For instance, consider the discrete points $p = (0, 0)$, $q = (30, 30)$ and $r = (10, 40)$, as illustrated in Figure 1.27. They correspond to the points p, q and r shown in Figure 1.26 with their coordinates multiplied by a factor 10. The previous contradiction still occurs since $d_{3,4}(p, q) = 120$, $d_{3,4}(p, r) = 130$, $d_{\mathrm{E}}(p, q) = 10\sqrt{18}$ and $d_{\mathrm{E}}(p, r) = 10\sqrt{17}$. Therefore, $d_{3,4}(p, q) < d_{3,4}(p, r)$ and $d_{\mathrm{E}}(p, r) > d_{\mathrm{E}}(p, q)$.

The points q and r are now mapped onto q' and r' via a counterclockwise rotation of $\frac{\pi}{8}$ around p. The coordinates of the rotated points are rounded for q' and r' to be the nearest integer points. This yields $q' = (16, 39)$ and $r' = (33, 25)$.

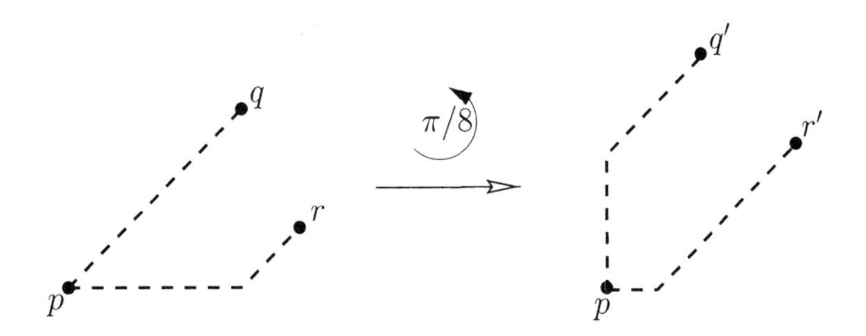

Figure 1.27 Non-invariance of $d_{a,b}$ against rotation

Remark 1.46:

It is important to note that, although a nearest integer truncation has been operated on the coordinates after the rotation, the order induced by d_E *on* p, q *and* r *is preserved (i.e.* $d_E(p,q) > d_E(p,r) \Rightarrow d_E(p,q') > d_E(p,r')$*).*

For this configuration, $d_{3,4}(p,q') = 133$, $d_{3,4}(p,r') = 124$, $d_E(p,q') \simeq 42.15$ and $d_E(p,r') \simeq 41.4$. Hence, $d_{3,4}(p,q') > d_{3,4}(p,r')$ and $d_E(p,q') > d_E(p,r')$. Therefore, for both d_E and $d_{3,4}$, the closest point to p is now r'. After rotation, the ordering induced by the discrete distance $d_{a,b}$ matches that of the Euclidean distance d_E and the closest point from p switches to r'.

It is often the case that actual distance values are only relevant for the ordering they induce between points. Discrete distances generally fail to satisfy the invariance of this ordering under rotation or similar geometric transformations.

Geometrically, this weakness can be explained as follows. Given D as a value of the discrete distance $d_{a,b}$ and $p = (0,0)$ as centre. Let us assume that there exist different points $q_i = (x_i, y_i)$ such that $d_{a,b}(p, q_i) = D$. In Figure 1.28(A), $a = 3$, $b = 4$, $D = a.b = 120$. Then $q_0 = (10b, 0) = (40, 0)$, $q_1 = (10a, 10a) = (30, 30)$.

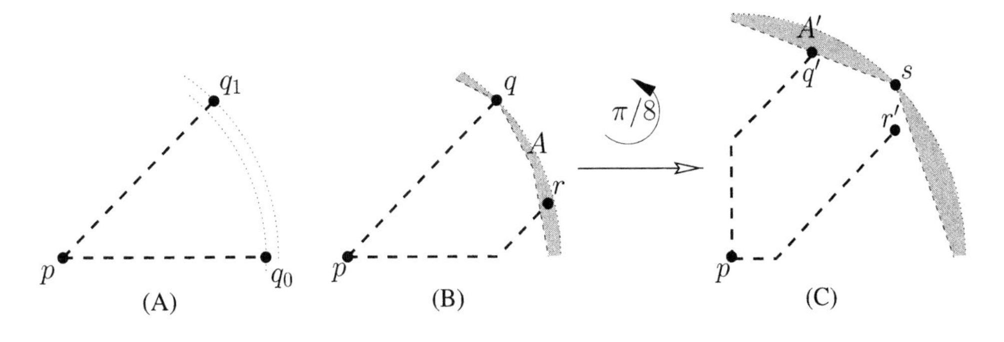

Figure 1.28 Geometrical explanation of the non-invariance of $d_{a,b}$

Each of the points q_i generally induces a different value for $d_E(p, q_i)$. Let R be the maximum of these values, $R = \max_i(d_E(p, q_i)$ and $q \in \{q_i\}$ such that

$d_E(p,q) = R$ and $d_{a,b}(p,q) = D$. In Figure 1.28(A), $R = 10a\sqrt{2} = 30\sqrt{2}$ and $q = q_1$. By definition of R, $\Delta_E(p,R)$ is the minimal Euclidean disc that contains all the points in $\Delta_{a,b}(p,D)$ ($\Delta_{a,b}(p,D) \subset \Delta_E(p,R)$). However, because of the distance approximation error, there exists a non-empty area A between the Euclidean and the discrete discs (i.e. $A = \Delta_E(p,R) - \Delta_{a,b}(p,D) \neq \emptyset$). Now, it is clear that for any integer point $r \in A$, $d_{a,b}(p,r) > d_{a,b}(p,q)$ and $d_E(p,r) < d_E(p,q)$ (see Figure 1.28). This case is precisely the example illustrated in Figure 1.27 where $D = 120$, $R = 10\sqrt{18}$ and $r = (40,10)$.

After rotation, $d_E(p,q')$ may not (and generally does not) represent the radius R' of the minimal Euclidean disc that contains the discrete disc $\Delta_{a,b}(p,D')$ with $D' = d_{a,b}(p,q')$. Continuing the example, after rotation of $\frac{\pi}{8}$ around p, the configuration shown in Figure 1.28(B) is mapped onto that shown in Figure 1.28(C). The point $s = (34,31)$ is such that $d_{3,4}(p,s) = D' = 133$ and leads to the minimal Euclidean disc $\Delta_E(p,R')$ that contains $\Delta_{a,b}(p,D')$ with $R' = d_E(p,s) \simeq 46.01$. The point $r' = (33,25)$ is no longer included in $A' = \Delta_E(p,R') - \Delta_{a,b}(p,D')$ and it is easy to verify that $d_{3,4}(p,q') > d_{3,4}(p,r')$ and $d_E(p,q') > d_E(p,r')$. In this case, the ordering induced by $d_{3,4}$ matches that of d_E.

The key to the non-invariance of discrete distances under geometric transformations lies in the decomposition of a discrete distance value D into different combinations of a- and b-moves. Detailing such decompositions and the characterisation of Euclidean discs that contain discrete discs, one can prove that the optimal move lengths in the 8-neighbourhood for the definition of $d_{a,b}$ are $a = 3$ and $b = 4$. Using such coefficients, the order of $d_{a,b}$ matches that of d_E up to the upper bound $R = \sqrt{17}$ (see [42, 92]). Similarly, it has been shown that the ordering of a discrete distance based on the 16-neighbourhood matches that of d_E up to a maximal bound of $R = \sqrt{104}$ for the optimal move lengths $a = 19$, $b = 27$ and $c = 42$ (see [42]).

Problems induced by the use of discrete distances are not detailed further at this stage. The topic of computing discrete and Euclidean distances will be revisited in Chapter 5 when dealing with distance transformations. Algorithmic solutions to these problems will then be presented.

Chapter 2

DISCRETE GEOMETRY

2.1 Introduction

Discrete geometry aims for a characterisation of the geometrical properties of a set of discrete points. Geometrical properties of a set are understood to be global properties. In discrete geometry, points are grouped, thus forming discrete objects, and it is the properties of these discrete objects that are under study. In contrast, digital topology as described in Chapter 1 allows for the study of the local properties between discrete points within such an object. In short, topological properties such as connectivity and neighbourhood are first used to define discrete objects and discrete geometry then characterises the properties of these discrete objects.

A possible approach is to consider discrete objects as digitisation of continuous objects and to map the properties of the original continuous object in the classic Euclidean geometry onto properties in discrete geometry. In this context, there is first a need for definitions and models of digitisation schemes. The resulting discrete geometrical properties will clearly be highly dependent on the digitisation scheme.

Ideally, one would rather seek discrete geometrical properties that are unrelated to the context in which the discrete object is studied (i.e. independent of the digitisation scheme used). Generally, a combined approach is used. Definitions of concepts in discrete geometry are given via relations to their continuous counterparts (e.g. straightness, convexity). However, characterisations of such properties are derived in a purely discrete approach. The aim of such characterisations is the development of algorithms related to geometric properties. For example, given a set of discrete points, one would be able to test it against a definition of discrete straightness (i.e. the fact that it is a digital straight segment or otherwise) without any reference to a specific continuous segment. A similar case arises for convexity. The characterisation for discrete convexity will be given without reference to any specific continuous convex set.

Characterisations of discrete geometrical properties which do not refer to any continuous object are important since algorithms corresponding to such properties should be based on local considerations only, as was the case for discrete distances (see Section 1.4).

Section 2.2 introduces and details the concept of straightness for a discrete set of points on square lattices (for the case of triangular lattice, see for example [45]). This concept is first related to its continuous counterpart via the definition of a simple digitisation scheme. Different characterisations of a digital straight segment in the 8-neighbourhood space are then given. Extensions of the characterisations of straightness in the particular cases of the 4- and 16-neighbourhood spaces are also presented. Furthermore, complementary results concerning efficient representations of the digitisation of a continuous segment are given.

Discrete convexity is presented in Section 2.3 as a natural extension of discrete straightness. On the square lattice, different definitions of discrete convexity have been given. Their inter-relationships are detailed in Section 2.3.2. Discrete convex hulls based on these definitions are also presented.

Remark 2.1:

Based on the argument developed in Section 1.4.4, triangular lattices will not be detailed here. For such a study, the interested reader is referred to [53].

Curvature is another important geometrical concept in Euclidean geometry. Its discrete counterpart results in the geometrical characterisation of discrete circular objects. Few advances have been made on this topic in discrete geometry. These are summarised in Section 2.4. Finally, for the sake of completeness, Section 2.5 briefly considers inter-relationships such as parallelism and orthogonality that can be characterised between discrete objects.

2.2 Discrete straightness

In the following, a real point is a point of the continuous space $\mathbb{R}^2$, a discrete point is an integer point on the square lattice (i.e. a point in the discrete space $\mathbb{Z}^2$). The continuous segment $[\alpha, \beta]$ is the part of the straight line (α, β) in $\mathbb{R}^2$ which passes by these two points.

In the continuous space $\mathbb{R}^2$, straightness can be characterised in different equivalent ways. Analytically, in the plane (x, y), a straight line is the representation of a linear function L which takes the form $L : y = \sigma x + \mu$, with $(\sigma, \mu) \in \mathbb{R}^2$, where σ is called the slope of the straight line. Geometrically, a continuous straight line is a continuous set of points such that any pair of real points within it defines the same slope.

In the discrete space, straightness is referred to as discrete straightness. The following introductory definition for this concept is given.

Definition 2.2: Digital straight segment

A discrete set of points is a digital straight segment if and only if it is the digitisation of at least one continuous straight segment.

Definition 2.2 only takes its full meaning when the digitisation scheme is defined. For this purpose, a simple scheme is first introduced. The aim here is not to achieve the study of a particular digitisation scheme but rather to create a consistent link between continuous and discrete spaces in order to introduce straightness in the discrete space (Section 2.2.1). For a detailed study of digitisation schemes, the reader is referred to Chapter 4.

2.2.1 Discrete straightness in the 8-neighbourhood space

2.2.1.1 Digitisation scheme

The most commonly used digitisation scheme is the grid-intersect quantisation [43] and is presented here via the following example. This scheme and its properties are further detailed in Section 4.1.3.

Consider the continuous segment $[\alpha, \beta]$ and the square lattice shown in Figure 2.1. The intersection points between $[\alpha, \beta]$ and the lattice lines are mapped onto their nearest integer points. In case of a tie, the discrete point which is locally at the left of $[\alpha, \beta]$ is selected ($[\alpha, \beta]$ is oriented from α to β). This digitisation scheme is illustrated in Figure 2.1 by the fact that intersections between $[\alpha, \beta]$ and lattice lines are mapped to their closest discrete points on the lattice (see Chapter 4 for a complete study of this digitisation scheme).

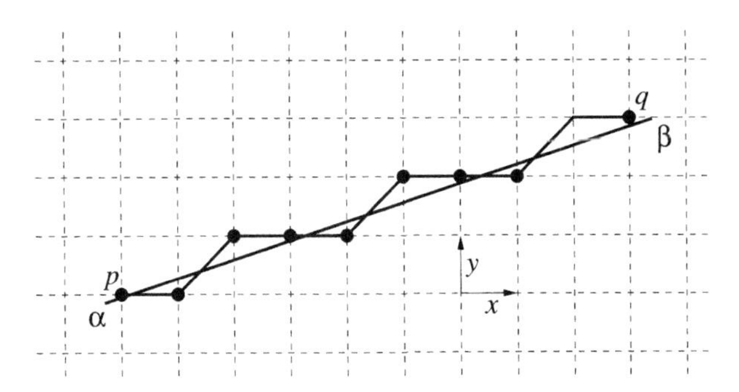

Figure 2.1 Grid-intersect quantisation

The set of discrete points $\{p_i\}_{i=0,\ldots,n}$ resulting from the digitisation of a continuous segment $[\alpha, \beta]$ is called the digitisation set of $[\alpha, \beta]$. Chapter 4 details the properties of the digitisation sets that result from the grid-intersect quantisation. In particular, it is shown that the grid-intersect quantisation of a continuous straight segment is an 8-digital arc (see Proposition 4.13).

The grid-intersect quantisation scheme creates a direct link between the definition of discrete straightness and that of straightness in Euclidean geometry. When defining a inverse process, one can characterise a set of discrete points $\{p_i\}_{i=0,\ldots,n}$ as a digital straight segment by defining a continuous segment $[\alpha, \beta]$ whose digitisation would yield $\{p_i\}_{i=0,\ldots,n}$ (see Section 4.1.3). However, this characterisation would not be efficient in practice since it does not allow for the use of integer arithmetic nor it does clearly suggest a precise algorithmic method for characterising discrete straightness. Moreover, such a characterisation is seen as global in the sense that the complete set of discrete points is to be related to a continuous segment. By contrast, local computation which would allow for the definition of an incremental algorithm (i.e. an algorithm considering successively each point and its neighbourhood only) is highly desirable. The next section addresses this problem in presenting characterisations for discrete straightness using discrete space properties only.

2.2.1.2 Freeman's codes

The first characterisation of a digital straight segment was given by Freeman [43]. This characterisation is descriptive and makes use of codes which are defined for all possible moves in the 8-neighbourhood. The particular structure of a sequence of such codes (i.e. the chain-code) is then used to characterise discrete straightness (Proposition 2.4).

Definition 2.3: Freeman's codes and chain-code

All possible moves in the 8-neighbourhood are numbered successively counter-clockwise from 0 to 7, as shown in Figure 2.2.

The encoding $\{c_i\}_{i=1,\ldots,n}$ *(*$c_i \in \{0, 1, \ldots, 7\}$*) of a given sequence of 8-moves defined by the discrete points* $\{p_i\}_{i=0,\ldots,n}$ *is called the chain-code of this sequence.*

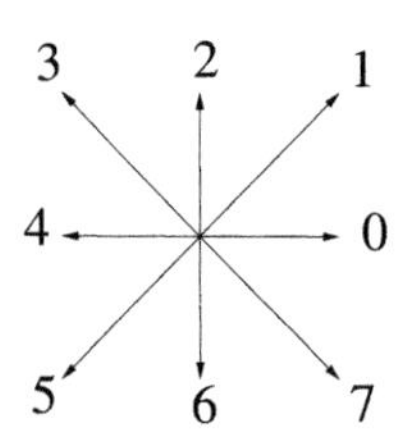

Figure 2.2 Freeman's codes in the 8-neighbourhood

Example: Chain-code

The chain-code of the 8-move sequence depicted in Figure 2.3 is

$$\{0, 0, 1, 3, 0, 0, 0, 6, 7, 0, 2, 2, 2, 4, 4, 4, 4, 4, 4, 4, 6, 6\}$$

Note that any 8-move sequence can be encoded in two ways, depending on the orientation chosen. For instance, if the other orientation is chosen in the example shown in Figure 2.3, the resulting chain-code is

$$\{2, 2, 0, 0, 0, 0, 0, 0, 0, 6, 6, 6, 4, 3, 6, 4, 4, 4, 7, 5, 4, 4\}$$

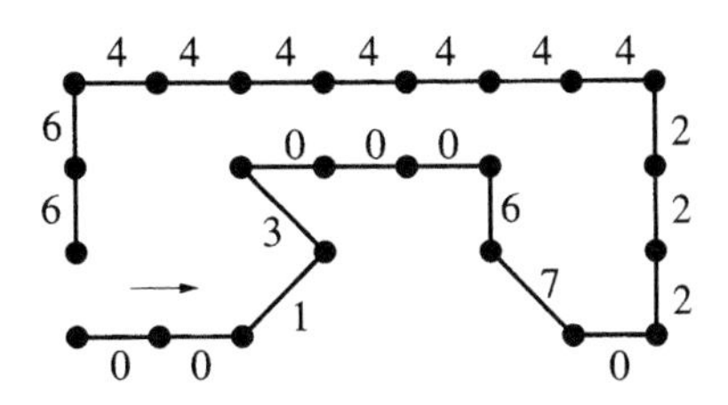

Figure 2.3 An example of the use of Freeman's codes

Nevertheless, there is a one-to-one relation between the two possible chain-codes of an 8-move sequence. The conversion rule of a chain-code $\{c_i^+\}_{i=1,\ldots,n}$ defined by a sequence of moves to the chain-code $\{c_i^-\}_{i=1,\ldots,n}$ of the same sequence with the opposite orientation is given by $c_i^- = (c_{n-i+1}^+ + 4) \bmod 8$, for all $i = 1, \ldots, n$.

There is no restriction on the fact that the 8-move sequence should define an 8-digital arc for it to yield a correct chain-code. Moreover, the choice of the starting point is arbitrary and defines an orientation for the encoded move sequence. Finally, it is clear that, given the chain-code and the position of the starting point of the 8-move sequence which is to be recovered, the positions of all other points are well-defined.

Characterisations of digital straight segment using chain-code sequences are formulated as in Proposition 2.4. Since the grid-intersect quantisation of a continuous straight segment is an 8-digital arc (see Proposition 4.13), Proposition 2.4 below assumes that the 8-move sequence considered forms an 8-digital arc.

Proposition 2.4:

An 8-digital arc is a digital straight segment if and only if its chain-code satisfies the following conditions [43]:

(i) At most two types of codes can be present, and these can differ only by unity, modulo eight.

(ii) One of the two code values always occurs singly.

(iii) Successive occurrences of the code occurring singly are as uniformly spaced as possible.

Example: Chain-code of a digital straight segment

Consider the 8-chain-code $\{0, 1, 0, 0, 1, 0, 0, 1, 0\}$ induced by the digital straight segment displayed in Figure 2.1 (oriented from α to β). Firstly, condition (i) is satisfied since only codes '0' and '1' are present in the chain-code. Typically, the two codes occurring in the chain-code of a digital straight segment are given by the slope of the continuous segment and the orientation defined during the encoding. In this case, the slope σ is such that $0 \leq \sigma \leq 1$. Considering the orientation, the slopes 0 and 1 correspond to the moves 0 and 1.

Moreover, code '1' occurs singly and, hence, condition (ii) is satisfied. Condition (iii) is rather vague. In this case one could say that it is satisfied since the code '0' appears in runs of lengths 1 and 2 and uniformly spaced within the chain-code. However, this criterion cannot be formally used in an automated process for characterising straightness. Therefore, this example shows that the chain-code $\{0, 1, 0, 0, 1, 0, 0, 1, 0\}$ induced by the digitisation of a continuous segment effectively satisfies the conditions given in Proposition 2.4 but does not enable one to formally characterise discrete straightness.

At this stage, Proposition 2.4 does not allow for the development of algorithms for testing discrete straightness. As mentioned in the above example, condition (iii) in Proposition 2.4 is not precise enough to formally differentiate a digital straight segment from a general digital arc. It has been proposed to replace condition (iii) from Proposition 2.4 by conditions (iii), (iv) and (v) in Proposition 2.5 (see [61,137]).

Proposition 2.5:

An 8-digital arc is a digital straight segment if and only if its chain-code satisfies the following conditions [61, 137]:

(i) At most two types of codes can be present, and these can differ only by unity, modulo eight.

(ii) One of the two code values always occurs singly.

(iii) The other code has run of only two lengths which are two consecutive integers.

(iv) One of the run lengths can occur only once at a time.

(v) For the run length that occurs in runs, these runs can themselves have only two lengths, which are consecutive integers, and so on.

Algorithms which test for straightness of digital arcs can be implemented from such propositions (e.g. see [61]). They are based on different rules derived from the conditions given in Propositions 2.4 and 2.5. Such algorithms will not be presented here and the interested reader is referred to the relevant literature (e.g. [43,61,137]).

2.2.1.3 Chord properties

This section introduces a different class of characterisations for discrete straightness, called *chord properties*. Originally proposed by Rosenfeld [137], the chord property (Proposition 2.6) remains one of the major results in discrete geometry. Variations and generalisations of the original characterisation have been proposed and are also presented in this section (see also Sections 2.2.2 and 2.2.3).

Chord properties are typically based on the following principle. Given a digital arc $P_{pq} = \{p_i\}_{i=0,\dots,n}$ from $p = p_0$ to $q = p_n$, the distance between continuous segments $[p_i, p_j]$ and the broken line $\bigcup_i [p_i, p_{i+1}]$ is measured via a discrete distance function and should not exceed a certain threshold (see Figure 2.4, where the shaded area represents the distance between P_{pq} and a continuous segment $[p_i, p_j]$).

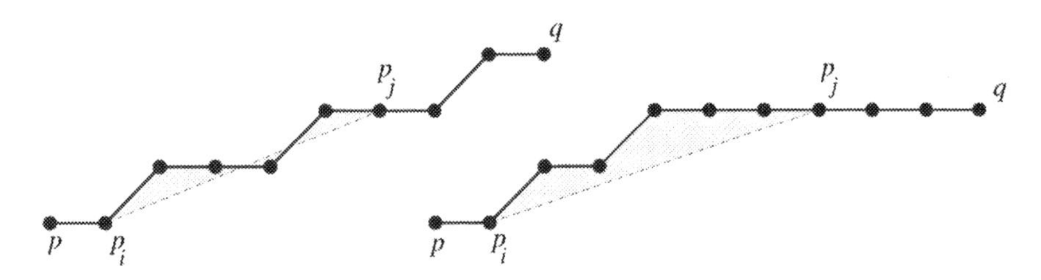

Figure 2.4 A general sketch for the chord properties

In other words, a polygon which surrounds the digital arc is defined and continuous segments defined between discrete points in the digital arc are to be contained in that polygon. Such a polygon will be referred to as a visibility polygon since the previous description suggests that a point should be visible from another within this polygon. Proposition 2.6 first introduces the chord property as originally formulated in [137].

Proposition 2.6:

An 8-digital arc $P_{pq} = \{p_i\}_{i=0,\dots,n}$ satisfies the chord property if and only if, for any two discrete points p_i and p_j in P_{pq} and for any real point α on the continuous segment $[p_i, p_j]$, there exists a discrete point $p_k \in P_{pq}$ such that $d_8(\alpha, p_k) < 1$.

Remark 2.7:

In Proposition 2.6, the definition of the (chessboard) d_8 distance is extended to real points via its analytical characterisation given by Proposition 1.29 (i.e. $d_8(p, q) = \max(|x_\beta - x_\alpha|, |y_\beta - y_\alpha|)$ for any $\alpha = (x_\alpha, y_\alpha)$ and $\beta = (x_\beta, y_\beta)$ in $\mathbb{R}^2$).

Geometrically, the chord property and the resulting visibility polygon can be illustrated by Figure 2.5. Given the digital arc P_{pq}, the shaded polygon in Figure 2.5 illustrates the set of points $\alpha \in \mathbb{R}^2$ such that there exists a discrete point $p_k \in P_{pq}$ such that $d_8(\alpha, p_k) < 1$ (i.e. the visibility polygon is the union of 8-discs of unit radii centred at every discrete point p_k of P_{pq}). From Proposition 2.6, P_{pq} satisfies the chord property if and only if the continuous segment $[p_i, p_j]$ is totally contained in this area for any i and j in $\{0, \ldots, n\}$. The chord property can therefore be reformulated as follows: "An 8-digital arc $P_{pq} = \{p_i\}_{i=0,\ldots,n}$ satisfies the chord property if and only if any point p_i is visible from any other point p_j within the visibility polygon defined by $\{\alpha \in \mathbb{R}^2$ such that $d_8(p_k, \alpha) < 1$ for all $k = 0, \ldots, n\}$".

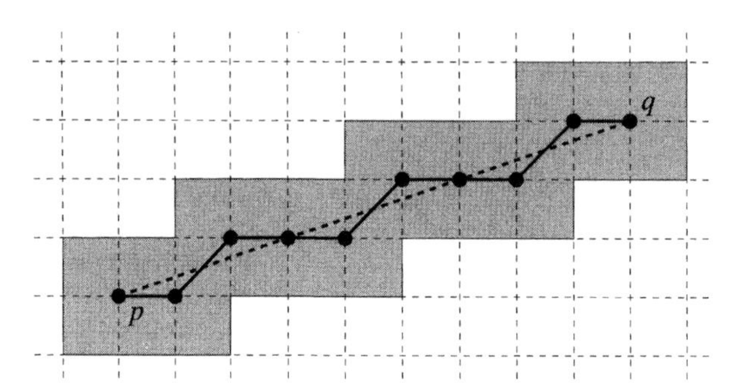

Figure 2.5 Example of the validity of the chord property

Figure 2.6 illustrates an instance where the conditions for the chord property are not satisfied. In this example, it is clear that $\alpha \in [p_1, p_8]$ is such that $d_8(p_k, \alpha) \geq 1$ for any $k = 0, \ldots, n$. In other words, α is outside the visibility polygon and p_1 is not visible from p_8 (and conversely) within the visibility polygon.

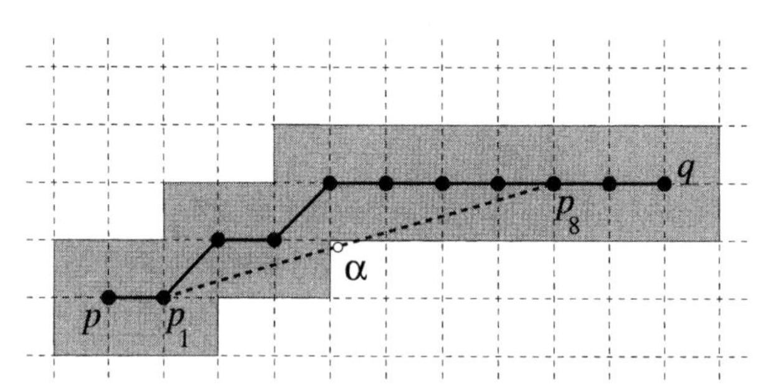

Figure 2.6 Example for the violation of the chord property

The chord property is an essential result since it provides an analytical formulation of discrete straightness via Theorem 2.8 below.

Theorem 2.8: [137]

In the 8-digital space:

(i) *The digitisation of a straight line is a digital arc and has the chord property.*

(ii) *If a digital arc has the chord property, it is the digitisation of a straight line segment.*

The original proof of Theorem 2.8 can be found in [137]. A simpler proof based on Santaló's theorem is given in [134].

Using this result, one can test for the discrete straightness of an 8-digital arc without reference to any related continuous segment whose digitisation would yield the digital arc in question. Moreover, the concept of visibility is important since it readily suggests a simple greedy algorithm which would successively test for the visibility of a point from other points in the digital arc under study.

Areas of visibility polygons defined in Proposition 2.6 are clearly not minimal. By definition, this polygon should be convex (see Section 2.3.1 for a definition of continuous convexity). The compact chord property aims for the reduction of visibility polygons by using the d_4 distance.

Proposition 2.9: [155]

An 8-digital arc $P_{pq} = \{p_i\}_{i=0,\ldots,n}$ satisfies the compact chord property if and only if, for any two distinct discrete points p_i and p_j in P_{pq} and for any real point α on the continuous segment $[p_i, p_j]$, there exists a real point $\beta \in \mathbb{R}^2$ in the broken line $\bigcup_i [p_i, p_{i+1}]$ such that $d_4(\alpha, \beta) < 1$.

Remark 2.10:

In Proposition 2.9, the definition of the d_4 distance is extended to real points via its analytical characterisation given by Proposition 1.27 (i.e. $d_4(\alpha, \beta) = |x_\beta - x_\alpha| + |y_\beta - y_\alpha|$ for all $\alpha = (x_\alpha, y_\alpha)$ and $\beta = (x_\beta, y_\beta)$ in $\mathbb{R}^2$).

The visibility polygon defined in the compact chord property is the set $\{\alpha \in \mathbb{R}^2$ such that $d_4(\beta, \alpha) < 1\}$ where $\beta \in \mathbb{R}^2$ is on the continuous segment $[p_i, p_j]$ for all $i, j = 0, \ldots, n$. It therefore corresponds to a unit 4-disc swept along the broken line $\bigcup_i [p_i, p_{i+1}]$.

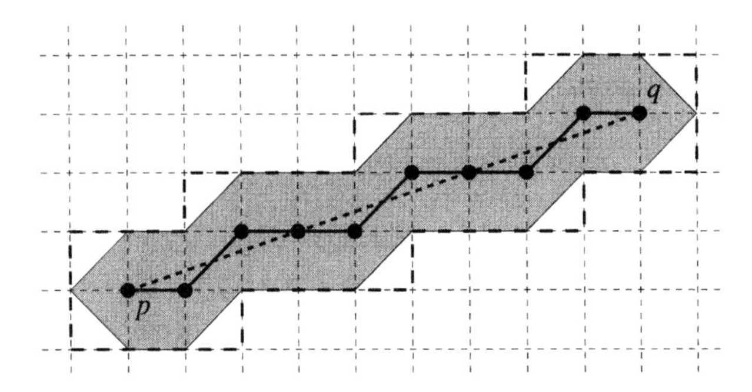

Figure 2.7 Example for the validity of the compact chord property

Figure 2.7 illustrates the difference between visibility polygons induced by the chord and compact chord properties, respectively. The shaded polygon is the visibility polygon defined by the compact chord property (Proposition 2.9), whereas the dashed bold polygon represents the contour of the visibility polygon defined by the chord property (Proposition 2.6). Since one is always included in the other, the term compact chord property was used. Using the same example as in Figure 2.6, Figure 2.8 shows that the digital arc also fails to satisfy the compact chord property. More generally, the exact equivalence between the chord and the compact chord properties is proved in [155].

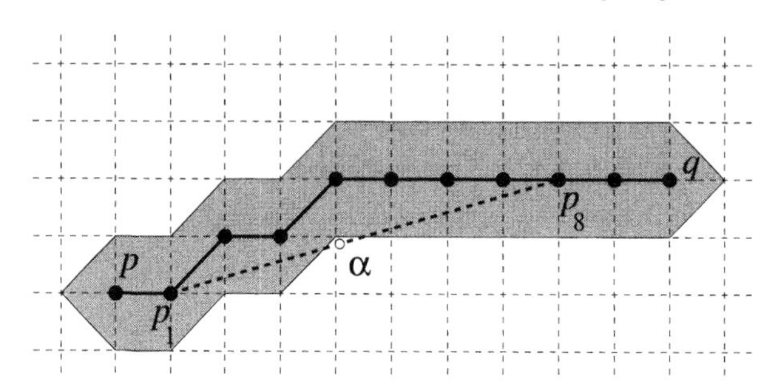

Figure 2.8 Example for the violation of the compact chord property

2.2.1.4 Upper and lower bounds of an 8-digital straight segment

Given two discrete points p and q, the grid-intersect quantisation of $[p, q]$ defines a particular digital straight segment between p and q. However, based on Freeman's codes and the definition of a shift operator (below), there exists a set of inter-related digital straight segments which can be defined between p and q [118]. This set defines a surface between p and q whose borders define the upper and lower bounds of the digital straight segment P_{pq}. In [118], an algorithm for constructing a digital straight segment from p to q is detailed and compared with that proposed by Bresenham [16].

Definition 2.11: Shift operator [118]

Given a chain-code sequence $\{c_i\}_{i=1,\dots,n}$, the shifted chain code is given by

$$\text{shift}(\{c_i\}) = \{c_2, c_3, \dots, c_n, c_1\}$$

Given the chain-code $\{c_i\}_{i=1,\dots,n}$ of a digital straight segment P_{pq}, the shift operator can be applied successively $n-1$ times on $\{c_i\}$ for generating $n-1$ shifted chain-codes, corresponding to different digital arcs from p to q. It is proved in [118] that any shifted chain-code defines a new digital straight segment from p to q. The union of all shifted digital straight segments forms an area which in turn defines the lower and upper bounds of the digital straight segment.

Example: Upper and lower bounds of a digital straight segment

Given the two discrete points p and q illustrated in Figure 2.9(A), an 8-digital straight segment can readily be obtained as the grid-intersect quantisation of the continuous segment $[p, q]$ (represented by a thick broken line). Its chain-code is given by $\{c_i\}_{i=1,\ldots,n} = \{0,1,0,0,1,0,0,1,0\}$. Consider the eight possible shifted chain-codes given in Table 2.1. These possible digital straight segments all lie within the shaded area associated with the continuous segment $[p, q]$. The upper and lower bounds of this area are called the upper and lower bounds of the digital straight segment P_{pq}. These bounds are represented as thick dotted and dashed lines, respectively, in Figure 2.9(B).

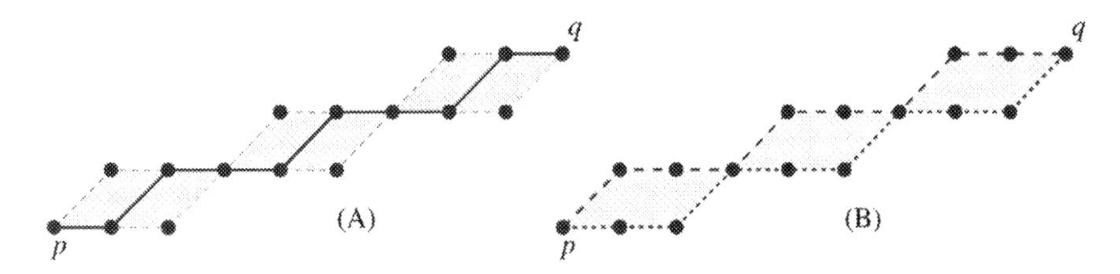

Figure 2.9 (A) Original digital straight segment and resulting shifted digital straight segments. (B) Upper (dashed) and lower (dotted) bounds of a digital straight segment

In this example, the upper and lower bound chain-codes are represented by shifts 1 and 2, respectively (see Table 2.1).

Shift	Chain-code
0	$\{0,1,0,0,1,0,0,1,0\}$[1]
1	$\{1,0,0,1,0,0,1,0,0\}$[2]
2	$\{0,0,1,0,0,1,0,0,1\}$[3]
3	$\{0,1,0,0,1,0,0,1,0\}$
4	$\{1,0,0,1,0,0,1,0,0\}$
5	$\{0,0,1,0,0,1,0,0,1\}$
6	$\{0,1,0,0,1,0,0,1,0\}$
7	$\{1,0,0,1,0,0,1,0,0\}$
8	$\{0,0,1,0,0,1,0,0,1\}$

[1]Grid-intersect quantisation.
[2]Upper bound.
[3]Lower bound.

Table 2.1 Shifted chain-codes

2.2.1.5 Parametrisation

After introducing different characterisations of discrete straightness, it is of interest to present results dealing with compact representations of 8-digital straight segments. These results assume that the given chain-code $\{c_i\}_{i=1,\ldots,n}$ is the digitisation of a continuous segment having a slope σ such that $0 \leq \sigma \leq 1$. In other words, it is assumed that the chain-code of the digital straight segment $\{c_i\}_{i=1,\ldots,n}$ consists of only 0's and 1's (other cases are equivalent by rotation or symmetry).

This first result was derived in [37] (see also [36]). In short, it states that the digitisation of a continuous segment can be uniquely represented by a set of four integers (n, q, p, s). In this set, n is the length of the chain-code (i.e. the cardinality of the digital straight segment represented) and q is its smallest periodicity (i.e. the smallest integer $k > 0$ such that $c_{i+k} = c_i$ for all i in $\{1, 2, \ldots, n-k\}$). p is the number of *b*-moves in the period (i.e. the repeated pattern) and s is the position of the first point in the period. Formulae for these parameters are given in Proposition 2.12 below.

Proposition 2.12: [37]

There is a one-to-one correspondence between an 8-digital straight segment chain-code $\{c_i\}_{i=1,\ldots,n}$ where $c_i \in \{0,1\}$ for all $i = 1, \ldots, n$ and the quadruple (n, q, p, s) defined as follows.

$$\begin{cases} n \text{ is the number of elements in } \{c_i\}_{i=1,\ldots,n} \\ \text{If } \{c_i\}_{i=1,\ldots,n} \text{ is aperiodic, then } q = n \\ \text{Else, } q = \min_k\{k \in \{1,2,\ldots,n\} \text{ such that } \forall i \in \{1,2,\ldots,n-k\}, c_{i+k} = c_i\} \\ p = \sum_{i=1}^{q} c_i \\ s \in \{0,1,2,\ldots,q-1\} \text{ is such that} \\ \qquad\qquad \forall i \in \{1,2,\ldots q\}, c_i = \left\lfloor \frac{p}{q}(i-s) \right\rfloor - \left\lfloor \frac{p}{q}(i-s-1) \right\rfloor \end{cases}$$

From Proposition 2.12, the chain code $\{c_i\}_{i=1,\ldots,n}$ of an 8-digital straight segment $\{p_i\}_{i=0,\ldots,n}$ characterised by the quadruple (n, q, p, s) can be obtained by the following equation

$$c_i = \left\lfloor \frac{p}{q}(i-s) \right\rfloor - \left\lfloor \frac{p}{q}(i-s-1) \right\rfloor, \qquad i = 1, \ldots, n$$

Example: Compact representation of a digital straight segment

Consider the 8-digital straight segment shown in Figure 2.5. It clearly belongs to the first octant since its chain-code $\{c_i\} = \{0,1,0,0,1,0,0,1,0\}$ is composed of 0's and 1's. The cardinality of this digital straight segment is $n = 9$. Moreover, one can easily see that $c_i = c_{i+3}$ for all $i = 1, \ldots, 5$. Therefore, $q = 3$. Within the chain-code period $\{0,1,0\}$, there is only one *b*-move. Hence, $p = 1$. The last parameter $s \in \{0, \ldots, q-1\} = \{0,1,2\}$ is not precisely defined. By elimination, one concludes that $s = 2$. It is easy to verify that, given the integer quadruple $(n = 9, q = 3, p = 1, s = 2)$, the complete chain-code can be retrieved using the reconstruction formula above. Therefore, the digital straight segment shown in Figure 2.5 can be uniquely represented by the integer quadruple $(9, 3, 1, 2)$.

Remark 2.13:

By definition, s is the only parameter which differentiates all possible digital straight segments between two points p and q. It can easily be seen in Figure 2.9(A) that the three possible digital straight segments between p and q whose union defines the upper and lower bounds of P_{pq} correspond to the values $s = 0$, $s = 1$ and $s = 2$, respectively.

As a further result, we also note that the authors of [37] derived analytical expressions for the set of all possible continuous segments whose digitisation could have generated the chain-code in question. Such a set of continuous segments is referred to as the domain of a digital straight segment [37, 79] (see Sections 2.5 and 4.1). Conversely, a characterisation of digital straightness in the 8-neighbourhood space can be defined by the fact that the chain-code of a general digital arc cannot be represented by such a quadruple of integers if it is not a digital straight segment. However, this characterisation does not clearly express digital straightness.

An equivalent representation of a straight line segment is given by the set of four integers (N, k, h, x_0) defined in Proposition 2.14 below.

Proposition 2.14: [85]

There is a one-to-one correspondence between the set of digital straight lines in the first octant, which start at a fixed point, and the set of integer parameters (N, k, h, x_0), where N is the cardinality of the digital straight segment and (k, h, x_0) are either $(N, N+1, 0)$ or satisfy

(i) $0 < h < k \leq N$

(ii) $\gcd(h, k) = 1$

(iii) $0 \leq x_0 \leq N - k$

The equivalence relationship between (n, q, p, s)- and (N, k, h, x_0)-representations is as follows

$$\left\{ \begin{array}{l} n = N \\ s = x_0 \\ \frac{p}{q} \text{ is the term immediately} \\ \quad \text{before } \frac{h}{k} \text{ in the Farey sequence} \\ \quad \text{of order } \max(N - x_0, x_0 + k) \\ \quad \text{if } \frac{h}{k} \leq 1 \\ \frac{p}{q} = 1 \text{ if } \frac{h}{k} > 1 \end{array} \right. \qquad \left\{ \begin{array}{l} N = n \\ x_0 = s \\ \frac{h}{k} \text{ is the unique term that succeeds } \frac{p}{q} \\ \quad \text{and satisfies } n - s - q < k \leq n - s \\ \quad \text{in some Farey sequence, if } \frac{p}{q} < 1 \\ \\ \frac{h}{k} = \frac{N+1}{N} \text{ if } \frac{p}{q} = 1 \end{array} \right.$$

Example: Compact representation of a digital straight segment

Continuing with the previous example, the representation of the digital straight segment shown in Figure 2.5 is $(N = 9, k = 5, h = 2, x_0 = 2)$. Conditions (i) to (iii) are clearly satisfied by this set of values. Moreover, the equivalence between (n, q, p, s) and (N, k, h, x_0) is satisfied since $\frac{h}{k} = \frac{2}{5}$ is the unique term succeeding $\frac{p}{q} = \frac{1}{3}$ in a Farey sequence that satisfies the condition $3 < k \leq 5$ (e.g. see Farey sequence of order 6 in the example given in Section 1.4.3). Conversely, $\frac{p}{q} = \frac{1}{3}$ is the term directly preceding $\frac{h}{k} = \frac{2}{5}$ in the Farey sequence of order $\max(N - x_0, x_0 + k) = 7$. Hence, for this digital straight segment, $(N, k, h, x_0) = (9, 5, 2, 2)$.

The advantage of (N, k, h, x_0)-representations is highlighted in [85], which gives an algorithm that computes these parameters for any given 8-digital straight segment which lies in the first octant. All four parameters (N, k, h, x_0) can be directly defined, whereas parameters (n, q, p, s) are typically found by elimination.

2.2.2 Straightness of 4-connected sets

In the case where the digitisation scheme used is not the grid-intersect quantisation, a 4-connected set of discrete point may be obtained as a digitisation of a continuous segment $[\alpha, \beta]$. Section 4.1.2 details the square-box quantisation in which a discrete point $r = (x_r, y_r)$ appears in the digitisation set of a straight segment $[\alpha, \beta]$ if and only if $[\alpha, \beta]$ intersects the open digitisation box associated with r, $]x_r - \frac{1}{2}, x_r + \frac{1}{2}[\times]y_r - \frac{1}{2}, y_r + \frac{1}{2}[$.

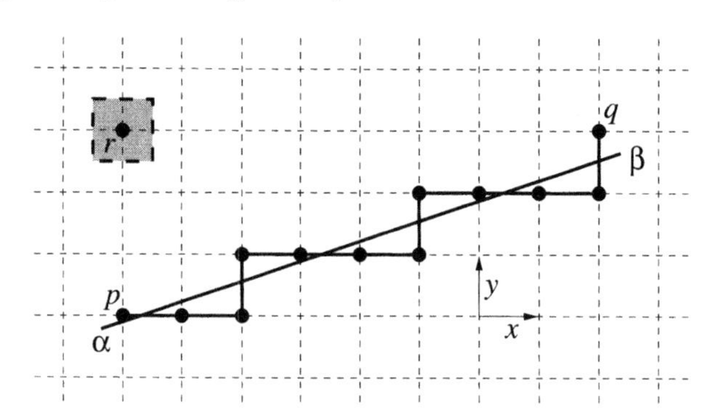

Figure 2.10 Square-box quantisation of the continuous segment $[\alpha, \beta]$ (the slope σ of $[\alpha, \beta]$ is such that $-1 < \sigma < 1$)

Figure 2.10 illustrates the square-box quantisation scheme. Any real point in the open digitisation box around the discrete point r is mapped onto r (the dashed borders highlight the fact that this continuous set is open). If a continuous segment $[\alpha, \beta]$ passes exactly in between two such boxes, then there exists at least one pair of 8-neighbouring discrete points ($s = (x_s, y_s)$ and $t = (x_t, y_t)$, say), such that the real point of coordinates $(\frac{x_s+x_t}{2}, \frac{y_s+y_t}{2})$ belongs to $[\alpha, \beta]$. In this case, one discrete point s or t is arbitrarily added to the square-box quantisation of $[\alpha, \beta]$. By convention, the discrete point which is locally at the left of $[\alpha, \beta]$ (oriented from α to β) is chosen to be in the square-box quantisation of $[\alpha, \beta]$.

In Chapter 4, it is shown that the resulting digitisation set $\{p_i\}_{i=0,\ldots,n}$ of a continuous straight segment $[\alpha, \beta]$ is a 4-arc (see Proposition 4.9).

The characterisation of discrete straightness in the 4-neighbourhood space was first introduced in [67]. The square-box quantisation of a continuous object is referred to as its cellular image and any 4-digital arc issued from this digitisation scheme is called a cellular arc. This terminology leads to the characterisation of cellular straight segments (i.e. 4-digital straight segments).

Proposition 2.15: [67]

A cellular arc (i.e. 4-digital arc) is the cellular image (i.e. the square-box quantisation) of a continuous straight segment if and only if its satisfies the chord property.

The visibility polygon defined around the 4-digital arc is therefore the visibility polygon induced by the chord property (e.g. see Figure 2.13 and Proposition 2.6).

Another characterisation of 4-discrete straightness is given in [135] using the strong chord property. This property relies on the definition of an alternative visibility polygon around the 4-arc in question.

Proposition 2.16: [135]

A 4-digital arc $P_{pq} = \{p_i\}_{i=0,\ldots,n}$ satisfies the strong chord property if and only if, for any two distinct discrete points p_i and p_j in P_{pq} and for any real point α on the continuous segment $[p_i, p_j]$, there exist two distinct discrete points p_k and p_l in P_{pq} such that p_k and p_l are 4-neighbours and $d_8(\alpha, p_k) + d_8(\alpha, p_l) < 2$.

Then, the following theorem characterises 4-discrete straightness.

Theorem 2.17: [135]

A 4-digital arc P_{pq} is a 4-digital straight segment if and only if it satisfies the strong chord property.

The visibility polygon induced by the strong chord property is therefore the union of sets $\{\alpha \in \mathbb{R}^2$ such that $d_8(\alpha, p_k) + d_8(\alpha, p_l) < 2\}$ for any two distinct p_k and p_l, 4-neighbours in P_{pq}. Figure 2.11(A) displays instances of such sets for the two possible configuration of 4-neighbours discrete points p_k and p_l. Figure 2.11(B) then displays the visibility polygon induced by the union of such sets positioned at each pair of 4-neighbours p_k and p_l in the 4-arc under study.

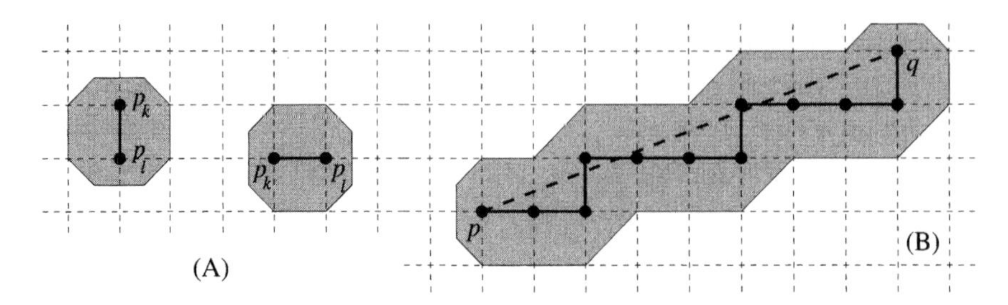

Figure 2.11 Strong chord property. (A) Basic sets which define the visibility polygon. (B) Visibility polygon associated with the 4-digital straight segment obtained from Figure 2.10

As a counterexample, Figure 2.12 displays an instance where the strong chord property is violated. The point $\alpha \in [p_2, q]$ is outside the visibility polygon. Therefore, the 4-arc P_{pq} is not a 4-digital straight segment.

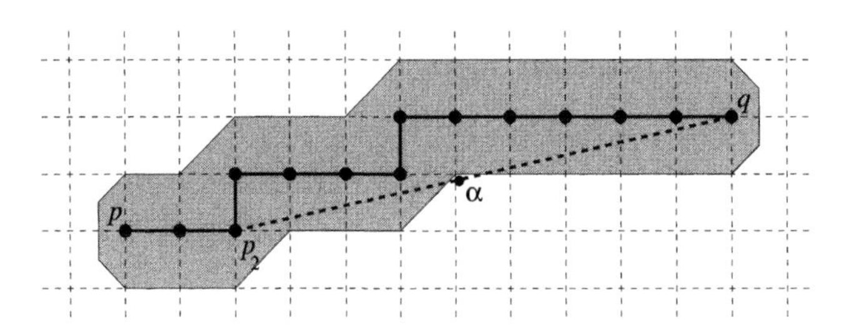

Figure 2.12 Violation of the strong chord property

Finally, Figures 2.13(A) and (B) compare visibility polygons defined by the chord and the strong chord properties for the sets of discrete points presented in Figures 2.11 and 2.12, respectively.

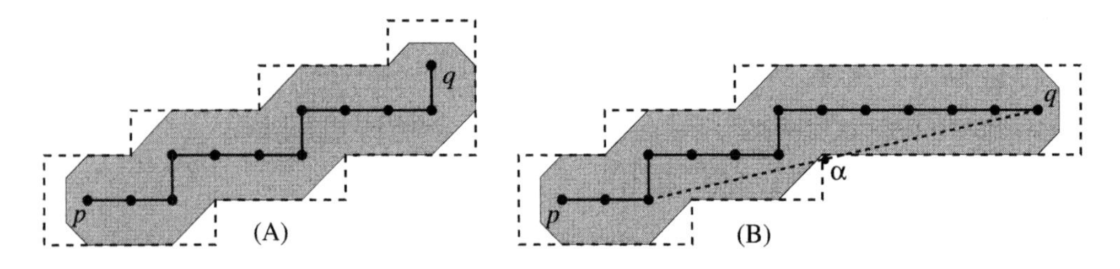

Figure 2.13 Comparison of visibility polygons defined by the chord and strong chord properties for the 4-connected set shown in (A) Figure 2.11 and (B) Figure 2.12

In both figures, the visibility polygon defined by the strong chord property is shown as a shaded area whereas the border of the visibility polygon defined by the chord property is shown as a dashed line.

2.2.3 Straightness of 16-connected sets

Characterising straightness in the 16-neighbourhood space is attractive since the digitisation of a continuous segment into a sequence of 16-moves is likely to be more precise than the digitisation of the same segment into an 8-move sequence. Moreover, redefining move codes from '0' to '15' (see Figure 2.14) would allow for a compact encoding via the use of the *knight*-moves (i.e. c-moves, see Chapter 1).

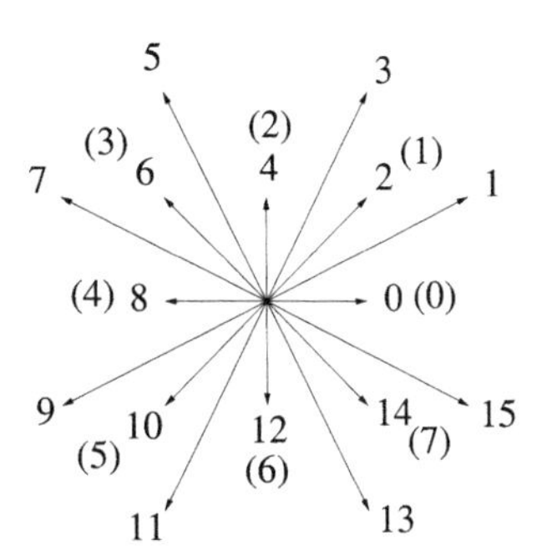

Figure 2.14 Move codes in the 16-neighbourhood. Equivalent Freeman's codes are shown in brackets when appropriate

Results concerning straightness in the 16-neighbourhood space are summarised in this section. These are typically formulated as extensions to the results in the 8-neighbourhood space. The work on 16-connected sets was first developed in [89]. Firstly, a one-to-one mapping from the 16- to the 8-neighbourhood space is defined in order to extend properties of the 8-neighbourhood space into the 16-neighbourhood space. The new discrete distance $d_{pq}(.,.)$ is then defined in the 16-neighbourhood space with the aim of formulating a property which is compatible with the chord properties defined in the 8-neighbourhood space for characterising digital straightness. The compact chord property is selected for this purpose and extended to the 16-compact chord property.

The following transformation defines a one-to-one mapping between the 8- and 16-neighbourhood spaces. 16-Moves defined in each octant are mapped onto 8-moves by the T_i transform defined in that octant. Because of the symmetry with respect to the origin, only four T_i transforms are required (see Figure 2.15). More formally, analytical definitions of T_i transforms (and their inverse T_i^{-1}) are given in Definition 2.18 below. Mappings induced from the 16-codes in each octant onto 8-codes (and their symmetric counterparts) are also included in this definition.

Definition 2.18: T_i transform

A transform T_i is defined in each octant and its symmetric counterpart with respect to the origin ($i = 0, \ldots, 3$). Each octant is represented as an interval of

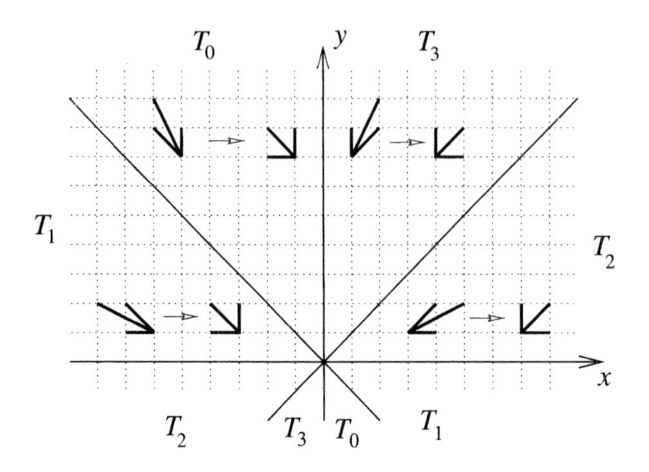

Figure 2.15 T_i transforms (numbered as in [89])

possible slopes σ for a continuous segment.

- $-\infty \leq \sigma \leq -1$: $T_0 : (x, y) \mapsto (x, y + x)$ *and* $T_0^{-1} : (x, y) \mapsto (x, y - x)$
  $T_0(4) = 2$, $T_0(5) = 3$ *and* $T_0(6) = 4$; $T_0(12) = 6$, $T_0(13) = 7$ *and* $T_0(14) = 0$
- $-1 \leq \sigma \leq 0$: $T_1 : (x, y) \mapsto (x + y, y)$ *and* $T_1^{-1} : (x, y) \mapsto (x - y, y)$
  $T_1(6) = 2$, $T_1(7) = 3$ *and* $T_1(8) = 4$; $T_1(14) = 6$, $T_1(15) = 7$ *and* $T_1(0) = 0$
- $0 \leq \sigma \leq 1$: $T_2 : (x, y) \mapsto (x - y, y)$ *and* $T_2^{-1} : (x, y) \mapsto (x + y, y)$
  $T_2(0) = 0$, $T_2(1) = 1$ *and* $T_2(2) = 2$; $T_2(8) = 4$, $T_2(9) = 5$ *and* $T_2(10) = 6$
- $1 \leq \sigma \leq +\infty$: $T_3 : (x, y) \mapsto (x, y - x)$ *and* $T_3^{-1} : (x, y) \mapsto (x, y + x)$
  $T_3(2) = 0$, $T_3(3) = 1$ *and* $T_3(4) = 2$; $T_3(10) = 4$, $T_3(11) = 5$ *and* $T_3(12) = 6$

Example: Mapping of a 16-digital arc onto an 8-digital arc using T_1

Consider the 16-digital arc P_{pq} defined by the 16-chain-code $\{8, 7, 8, 8, 7, 7\}$, as illustrated in Figure 2.16. The slope σ of $[p, q]$ is such that $-1 \leq \sigma \leq 0$. Hence, the T_1 and T_1^{-1} transforms are used.

Its mapping via the T_1 transform yields the 8-chain-code $\{4, 3, 4, 4, 3, 3\}$ which in turn defines the 8-digital arc $P_{p'q'}$.

The aim of the T_i transforms is to map a 16-digital arc onto an 8-digital arc. Since they are one-to-one mappings, T_i transforms can also be used for the definition of digital straightness in the 16-neighbourhood space. By Proposition 2.4, an 8-digital straight segment P_{pq} is an 8-digital arc composed of at most two moves whose codes differ by one modulo eight. Such an 8-digital arc allows for the use of one inverse transform only (see Definition 2.18 and Figure 2.15), which is denoted $T_{i^*}^{-1}$. By continuity of the transformation, $T_{i^*}^{-1}(P_{pq})$ is a 16-digital

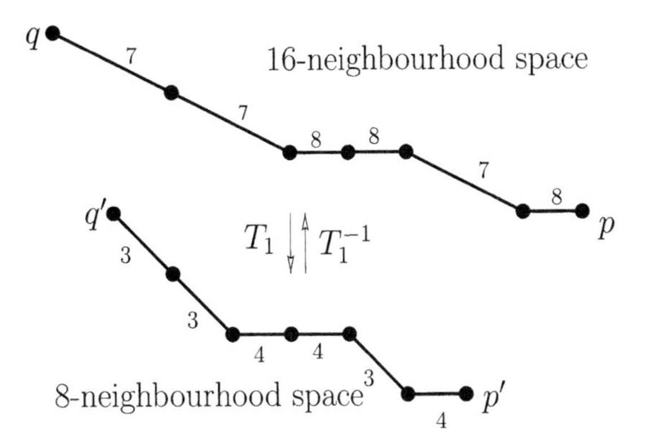

Figure 2.16 Example of T_1-mapping of a 16-digital arc onto an 8-digital arc

arc composed of at most two moves whose codes differ by one modulo 16. Such a 16-digital arc allows for the use of one transform only, namely T_{i^*}. Digital straightness can then be defined as in Definition 2.19 below.

Definition 2.19: 16-Digital straightness

A 16-digital arc is a 16-digital straight segment if and only if its mapping via the appropriate T_{i^*} transform satisfies the chord properties (i.e. is an 8-digital straight segment).

An analytical characterisation of 8-digital straightness was defined in Proposition 2.9 as the compact chord property. This result is to be extended to the 16-neighbourhood space via the definition of a new distance function which consists of the mapping of the d_4 distance via the T_i transform. Because of the dependence of the T_i transform on the slope of the 16-digital arc under study, the new distance will also show this dependence. For emphasising this dependence in the notation, the distance is denoted as d_{pq} where p and q define the octant which, in turn, defines the T_i transform.

Definition 2.20: d_{pq} distance in the 16-neighbourhood space

Given two discrete points $p = (x_q, y_p)$ and $q = (x_q, y_p)$, $d_i(p,q)$, $i = 0, \ldots, 3$ is defined as

$$\begin{aligned}
d_0(p,q) &= |\delta_{\mathrm{x}}(p,q)| + |\delta_{\mathrm{x}}(p,q) + \delta_{\mathrm{y}}(p,q)| \\
d_1(p,q) &= |\delta_{\mathrm{y}}(p,q)| + |\delta_{\mathrm{x}}(p,q) + \delta_{\mathrm{y}}(p,q)| \\
d_2(p,q) &= |\delta_{\mathrm{y}}(p,q)| + |\delta_{\mathrm{x}}(p,q) - \delta_{\mathrm{y}}(p,q)| \\
d_3(p,q) &= |\delta_{\mathrm{x}}(p,q)| + |\delta_{\mathrm{x}}(p,q) - \delta_{\mathrm{y}}(p,q)|
\end{aligned}$$

where $\delta_{\mathrm{x}}(p,q) = x_q - x_q$ and $\delta_{\mathrm{y}}(p,q) = y_q - y_q$.

Then, given two discrete points r and s, the value $d_{pq}(r,s)$ is defined as

$$d_{pq}(r,s) = d_{i^*}(r,s) \text{ with } i^* \text{ such that } d_{i^*}(p,q) = \min_{i=0,\ldots,3} d_i(p,q)$$

The newly defined distance d_{pq} is proved to be a distance in the discrete space since it satisfies the metric conditions given by Definition 1.15 (see [89]). Moreover, the value of i^* corresponds to the octant pairs defining the transformation T_i (see Figure 2.15).

In this context, an extension of the compact chord property, the 16-compact chord property, which allows for the characterisation of 16-straightness, can now be formulated.

Proposition 2.21: [89]

A 16-digital arc $P_{pq} = \{p_i\}_{i=0,\ldots,n}$ satisfies the 16-compact chord property if and only if, for any two distinct discrete points p_i and p_j in P_{pq} and for any real point α on the continuous segment $[p_i, p_j]$, there exists a real point $\beta \in \mathbb{R}^2$ in the broken line $\bigcup_i [p_i, p_{i+1}]$ such that $d_{pq}(\alpha, \beta) < 1$.

The following theorem concludes the study of 16-straightness.

Theorem 2.22: [89]

A 16-digital arc P_{pq} is a 16-digital straight segment as defined by Definition 2.19 if and only if it satisfies the 16-compact chord property.

Proof:
Immediate by definition of d_{pq}. If a 16-digital arc satisfies the 16-compact chord property, based on the T_i transform and the 16-distance d_{pq}, the 8-digital arc $T_{i^*}(P_{pq})$ satisfies the compact chord property. □

The extension of the compact chord property defines a visibility polygon in the 16-neighbourhood. Proposition 2.21 implies that, given any pair of discrete points (p_i, p_j) within a 16-digital straight segment P_{pq}, p_i (respectively p_j) is visible from p_j (respectively p_i) within this polygon. The visibility polygon is defined using the d_{pq} distance whose unit disc varies according to the slope of the real segment $[p, q]$.

Figure 2.17(A) displays the unit d_{pq} bowl defined in the same octant as T_2. By definition of T_2^{-1}, this unit bowl is mapped onto a d_4 unit disc by T_2. As for the compact chord property, the d_{pq} unit disc is swept along the 16-digital arc to define the 16-compact chord property visibility polygon. Figure 2.17(B) illustrates an example of a 16-digital arc satisfying the 16-compact chord property and its mapping via the T_2 transform (i.e. $i^* = 2$). By definition of T_2 and d_{pq}, if P_{pq} satisfies the 16-compact chord property, then $T_2(P_{pq})$ satisfies the compact chord property (and conversely, via T_2^{-1}). This is the case in Figure 2.17(B) where the visibility polygons are displayed as shaded areas.

Similarly, Figure 2.17(C) displays a case where P_{pq} does not satisfy the 16-compact chord property. The real point $\alpha \in [p_1, p_8]$ is outside the visibility polygon (shaded area). The use of the T_2 transform shows that the 8-digital arc $P_{p'q'} = T_2(P_{pq})$ does not satisfy the compact chord property. The real point $\alpha' = T_2(\alpha) \in [p'_1, p'_8] = T_2([p_1, p_8])$ is outside the visibility polygon defined from the compact chord property.

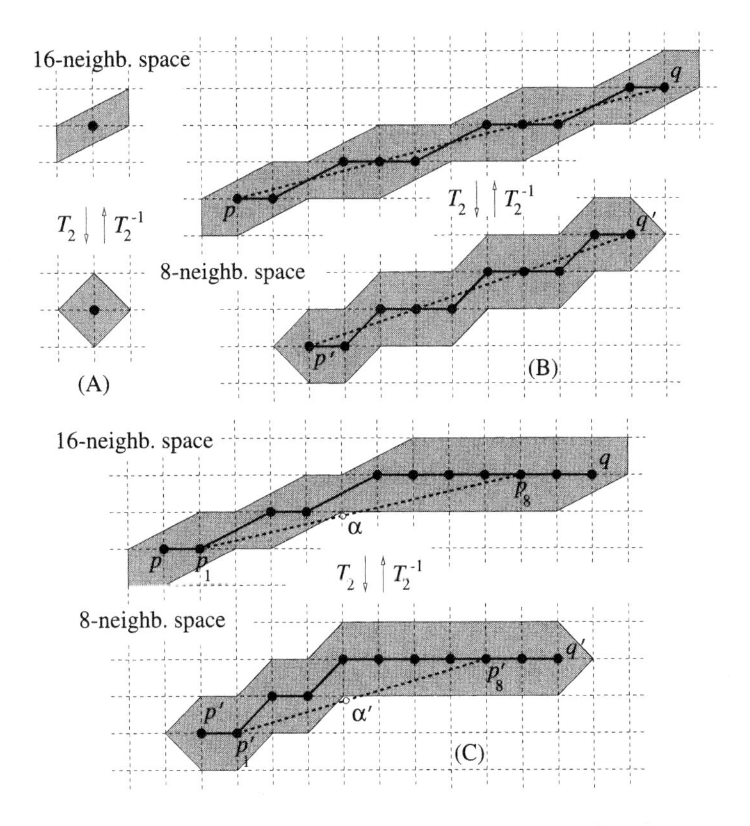

Figure 2.17 Example of verifications of the 16-compact chord property. (A) d_{pq} and d_4 unit discs. (B) P_{pq} (respectively, $P_{p'q'}$) satisfies the 16-compact chord property (respectively, compact chord property). (C) P_{pq} (respectively, $P_{p'q'}$) violates the 16-compact chord property (respectively, compact chord property)

In summary, one can take two approaches to check the straightness of a 16-digital digital arc $P_{pq} = \{p_i\}_{i=0,\ldots,n}$. Using the slope of the real segment $[p, q]$, one can uniquely define the appropriate $i^* \in \{0, \ldots, 3\}$ corresponding to the octant in which the digital arc lies. From then on, d_{pq} (and hence the associated visibility polygon) is defined. Then, in a similar fashion as in the 8-neighbourhood space, a simple greedy algorithm that checks visibility between pair of points (p_i, p_j) within this polygon can be tested.

The second approach uses the duality induced by the T_i transform. The appropriate T_{i^*} transform is uniquely defined using the slope of the real segment $[p, q]$. Then, as shown in Figure 2.17 where $i^* = 2$, checking for the (16-)straightness of the 16-digital arc P_{pq} is exactly equivalent to checking the (8-)straightness of the 8-digital arc $P_{p'q'} = T_{i^*}(P_{pq})$.

Remark 2.23:

More generally, the two approaches described above can readily be used for extending other properties from the 8-neighbourhood space into the 16-neighbourhood space. The duality created by the T_i transforms allows for taking one of the two approaches. One could map a property from the 8-neighbourhood space onto the 16-neighbourhood space and hence study the 16-digital object precisely. Alternatively, the 16-digital object could be mapped onto an equivalent 8-digital object on which 8-properties can be studied (see [89] for more details).

2.3 Discrete convexity

Convexity is an important shape descriptor in the continuous space. Many properties have been derived which ease the geometrical analysis of problems by simplifying their formulations. In this section, the concept of convexity is extended to the discrete space represented on the square lattice.

Convexity of discrete objects (i.e. sets of discrete points) will be referred to as discrete convexity. Different definitions for discrete convexity have been given. Equivalence between them has also been proved. A survey and analysis of the most important definitions of discrete convexity and their equivalence is presented here (see also [136] for a list of references on convexity).

2.3.1 A note on continuous convexity

The definition and properties related to convexity in the continuous plane $\mathbb{R}^2$ (i.e. continuous convexity) are first recalled. Details concerning these results can be found in most of introductory books on Euclidean and computational geometry (see, for example, [40, 120, 160]).

Definition 2.24: Continuous convexity

A set of points $S \subset \mathbb{R}^2$ is said to be convex if, for any two real points $\alpha = (x_\alpha, y_\alpha)$ and $\beta = (x_\beta, y_\beta)$ in S and any real number $\lambda \in [0, 1]$ the real point of coordinates $(\lambda x_\alpha + (1 - \lambda)x_\beta, \lambda y_\alpha + (1 - \lambda)y_\beta)$ belongs to S.

Remark 2.25:

Points in $\mathbb{R}^2$ can be identified with vectors from the origin. In the previous definition, the real point of coordinates $(\lambda x_\alpha + (1 - \lambda)x_\beta, \lambda y_\alpha + (1 - \lambda)y_\beta)$ can *therefore be denoted as $\lambda\alpha + (1-\lambda)\beta$. In other words, this point belongs to the continuous segment $[\alpha, \beta]$ if $\lambda \in [0,1]$.*

Hence, the previous definition is equivalent to the geometrical characterisation of continuous convexity below.

Proposition 2.26:

A non-empty set $S \subset \mathbb{R}^2$ is convex if and only if $\forall\, \alpha, \beta \in S$, $[\alpha, \beta] \subseteq S$.

Proof:
Immediate from Definition 2.24 and Remark 2.25. □

It follows from Proposition 2.26 that the intersection of two convex sets is convex. By contrast, the union of two convex sets is, in general, not convex. Given a set S, one can associate with it its convex hull.

Definition 2.27: Continuous convex hull

The convex hull of a continuous set S is the minimal convex set that contains S and is denoted $[S]$. Equivalently, the convex hull $[S]$ of the set S of real points is the intersection of all convex sets that contain S.

Remark 2.28:

The convex hull of a continuous convex set is equal to the set itself (e.g. see Figure 2.18(A)).

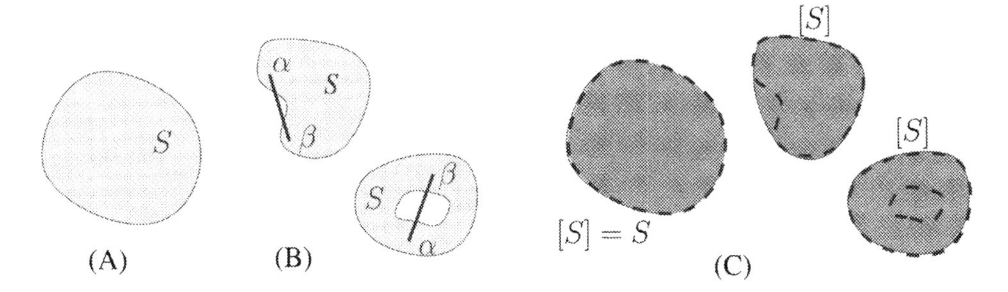

Figure 2.18 Examples of continuous (A) convex and (B) non-convex sets and (C) their respective convex hulls

Figures 2.18(A) and 2.18(B) illustrate examples of continuous convex and non-convex sets, respectively. In Figure 2.18(B), an example of a segment $[\alpha, \beta]$ which is not totally included in S is shown for each non-convex set S. Figure 2.18(C) shows the associated convex hulls of the same sets as in Figures 2.18(A) and 2.18(B) as dark shaded areas.

2.3.2 Discrete convexity

Results presented in this section summarise the development of definitions and characterisation of discrete convexity. Most of these results are presented in great detail in references [65, 66, 69, 70, 89, 132, 135]. We adopt the notation in reference [132] for introducing discrete convexity.

Definition 2.29: Notation for discrete convexity

- *Given a set of discrete points P, the cardinality of P is the number of discrete points that is included in P and is denoted $|P|$.*
- *Given a set of discrete points $P = \{p_i\}$, $\langle P \rangle$ is the set of discrete points contained in $[P]$, the continuous convex hull of P.*
- *If P contains a finite number of discrete points (i.e. $n = |P|$ is finite), then P can be written as $P = \{p_0, p_1, \ldots, p_n\}$. In this case, $\langle P \rangle$ can be equivalently written as $\langle p_0, p_1, \ldots, p_n \rangle$.*

The main definition of discrete convexity is referred to as cellular convexity [65, 66, 69, 70] and clearly refers to the square-box quantisation introduced in Section 2.2.2 (see Section 4.1.2 for a detailed study of this digitisation scheme).

Given a simply connected set of discrete points $P = \{p_i\}_{i=0,\ldots,n}$, with every discrete point $p_i = (x_i, y_i)$ is associated a unit digitisation box $B_i \subset \mathbb{R}^2$ centred at that point. Thus, $B_i = [x_i - \frac{1}{2}, x_i + \frac{1}{2}] \times [y_i - \frac{1}{2}, y_i + \frac{1}{2}]$. Let $S = \bigcup_i B_i$ be the set of real points included in the union of all sets B_i associated with P and let ∂S be the boundary of S. Given two discrete points p_i and p_j in P, $\mathcal{P}(P; p_i, p_j)$ is the set of polygons whose boundaries are part of the continuous segment $[p_i, p_j]$ and ∂S (see Figure 2.19). Using this notation, discrete convexity can then be defined as follows.

Definition 2.30: Discrete convexity (cellular convexity [65, 69])

A set of discrete points P is discrete convex if and only if for any two discrete points p_i and p_j in P, all discrete points contained in $\mathcal{P}(P; p_i, p_j)$ are also contained in P.

In Figure 2.19(A), discrete points which belong to P are displayed as black dots ($\bullet$). The dashed line around the digitisation boxes associated with discrete points in P represents ∂S. The shaded area illustrates an example of $\mathcal{P}(P; p_i, p_j)$. This area contains two discrete points s and t (displayed as empty discs ($\circ$)) which do not belong to P. Therefore, P is not convex. Using the same notation, Figure 2.19 illustrates a case where P is convex. In this case, for any pair of discrete points p_i, p_j in P, the area $\mathcal{P}(P; p_i, p_j)$ does not contain any discrete point which belong to P.

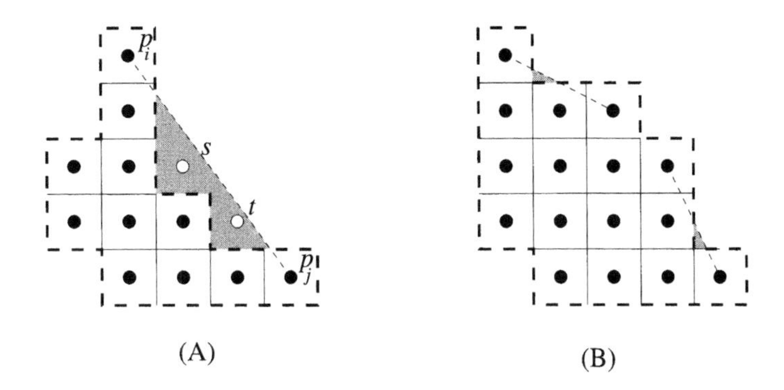

Figure 2.19 Definition of discrete convexity

Relating discrete convexity with continuous convexity can be done with the aid of Remarks 2.25 and 2.28. In short, these remarks state that a continuous set S is convex if and only if $[S] = S$. A characterisation of discrete convexity via the continuous convex hull of P can be formulated as follows.

Proposition 2.31: [65, 69]

A set of discrete points P is discrete convex if and only if any discrete point contained in the convex hull of P belongs to P. In short, P is discrete convex if and only if $\langle P \rangle = P$.

If P is a simple 8-connected set (see Definition 1.7 in Chapter 1), the equivalence between Proposition 2.31 and Definition 2.30 is obtained [65] (e.g. compare Figures 2.19 and 2.20). Figures 2.20(A) and 2.20(B) display the resulting characterisation of discrete convexity for the same sets P as in Figure 2.19. Clearly, in Figure 2.20(A), the convexity criterion fails for the same reason as it did using Definition 2.30.

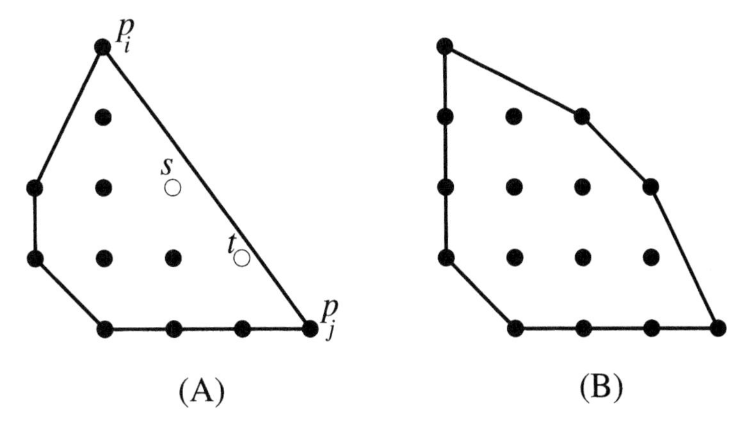

Figure 2.20 An equivalent characterisation of discrete convexity (to be contrasted with Figure 2.19)

Moreover, Proposition 2.31 holds if and only if for any discrete points p, q and r in P, $\langle p, q, r\rangle \subseteq P$ (see [132]). For a set of discrete points, such a property is called triangle-convexity. More formally, Definition 2.32 presents alternative definitions for discrete convexity.

Definition 2.32: [132]

Let P be a set of discrete points such that $|P|$ is finite.

(i) P is line-convex (or L-convex, for short) if $\forall\, p, q \in P$, $\langle p, q\rangle \subseteq P$.

(ii) P is triangle-convex (or T-convex, for short) if and only if $\forall\, p, q, r \in P$, $\langle p, q, r\rangle \subseteq P$.

(iii) Let $k = 4$ or 8. P is k-convex if and only if P is k-connected and L-convex.

Note that if a set is T-convex, then it is clearly L-convex. Conversely, 8-connectivity is required for an L-convex set to be T-convex (see Figure 2.21 for a counterexample).

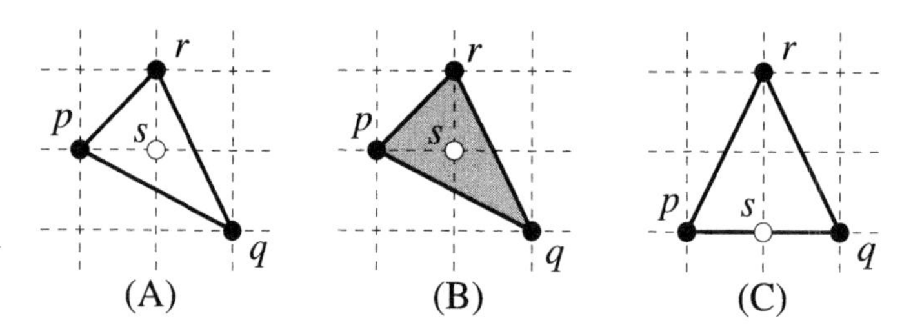

Figure 2.21 Relation between L- and T-convexity. (A) $\{p, q, r\}$ is L-convex but (B) not T-convex. (C) $\{p, q, r\}$ is not L-convex

In Figure 2.21(A) and (B), points p, q and r do not form an 8-connected component. When checking for L-convexity of $\{p, q, r\}$, one obtains the diagram displayed in Figure 2.21(A). This is a special case where $P = \bigcup\{\langle p, q\rangle \;\; \forall\, p, q \in P\}$. Therefore the set $\{p, q, r\}$ is L-convex. Figure 2.21(B) shows that the set $\{p, q, r\}$ is not T-convex, since s belongs to $\langle P\rangle = \langle p, q, r\rangle = \{p, q, r, s\}$ but not to $P = \{p, q, r\}$. In fact, it is easily seen that the T-convexity criterion fails for points whose addition would make P an 8-connected component (point s in this case). Similarly, Figure 2.21(C) displays an example where $\{p, q, r\}$ is not L-convex. The point s is included in $\langle p, q\rangle$ but not in P.

An 8-convex set is L-convex and 8-connected. Therefore, it is T-convex. Similarly, a 4-convex set is 4-connected and L-convex. Since the 4-neighbourhood of a point is included in its 8-neighbourhood, 4-connectivity implies 8-connectivity. Hence, 4-convexity is equivalent to L-convexity "plus" 8-connectivity, which, in turn, is equivalent to T-convexity. Consequently, the following proposition holds.

Proposition 2.33: [132]

Let P be a set of discrete points.

(i) If P is T-convex, then P is L-convex.

(ii) If P is L-convex and P is an 8-connected component, then P is T-convex.

L-convexity is presented in [95] as a definition for discrete convexity. Based on the above remarks and propositions, one can readily prove the following proposition.

Proposition 2.34: [65]

Given an L-convex set of discrete points P, if P is simply 8-connected (i.e. P does not contain any hole), then P is cellular convex as defined in Definition 2.30.

Figure 2.22 summarises as a chart relationships between the discrete convexity definitions introduced.

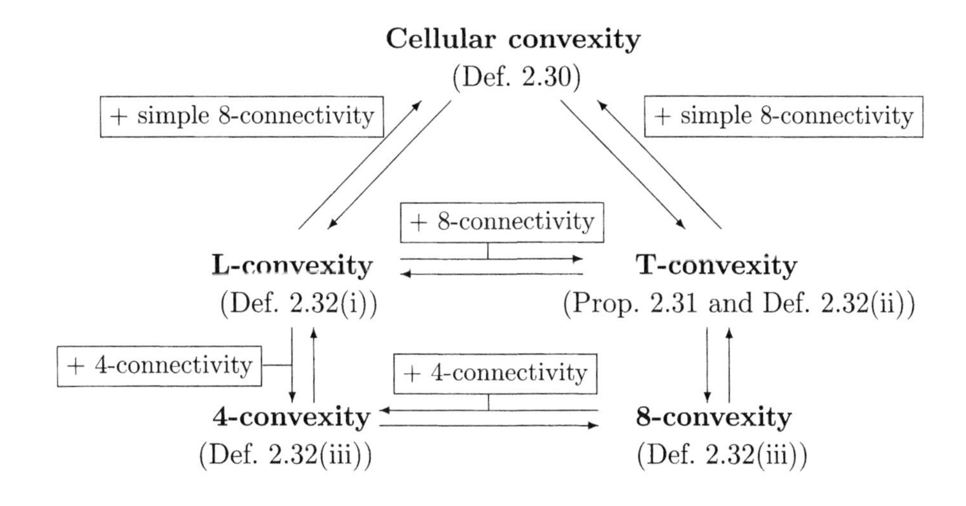

Figure 2.22 Relation between definitions of discrete convexity

Clearly, under the assumption that a set is simply 8-connected, all definitions of convexity for 8-connected sets are equivalent. 4-Convexity can be seen as a special case since 8-connectivity is the most used connectivity relationship in digital image processing. Moreover, by analogy with the continuous case, it makes sense to restrict the study of convexity of discrete sets to simply connected sets only. Therefore, discrete convexity will, in general, refer to any one of these definitions.

A geometrical characterisation of discrete convexity similar to that of continuous convexity (Proposition 2.26) can now be formulated.

Proposition 2.35: [69]

A set of discrete points P is discrete convex if and only if any point $p_i \in P$ is connected to any other point $p_j \in P$ by an 8-digital straight segment whose points belong to P.

Note that, by definition, an 8-digital arc is a simple 8-connected component. In that case, all definitions of discrete convexity presented above are equivalent. Conversely, a characterisation of discrete straightness can be formulated using discrete convexity.

Proposition 2.36: [69]

A digital arc is a digital straight segment if and only if it is discrete convex.

The result in Proposition 2.36 clearly applies to the 8-neighbourhood space. It can readily be extended to the 4-neighbourhood space using the strong chord property [135] (see Section 2.2.2) and to the 16-neighbourhood space using the 16-compact chord property [89] (see Section 2.2.3).

2.3.2.1 Discrete convex hull

In general, the definition of the discrete convex hull of a set of discrete points P is based on that of the continuous convex hull given by Definition 2.27.

Definition 2.37: Discrete convex hull

The discrete convex hull of a set of discrete points P is the intersection of all discrete convex sets that contain P.

As noted in [22, 132], 4- or 8-connectivity is not preserved by the intersection of discrete convex sets. In other words, although the discrete set resulting from the intersection of two discrete convex sets is discrete convex, it might not be connected.

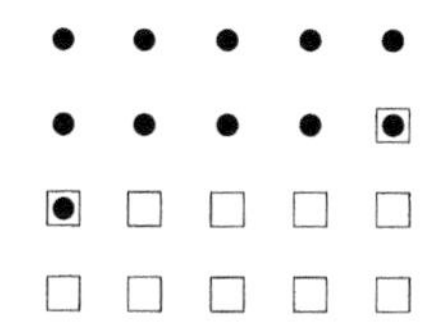

Figure 2.23 A special case of intersection for two discrete convex sets

Figure 2.23 shows an example where the intersection of two 8-connected discrete sets, represented by points • and □, respectively is not an 8-connected component.

In [65], convex hull simply refers to the continuous convex hull of the set of discrete points P in question and is called the Minimum Perimeter Polygon

(MPP). $\langle P \rangle$ is therefore the underlying definition for the discrete convex hull in this case. Geometrical definitions of a discrete convex hull generally follow this approach and prove to be mostly equivalent to the characterisation of $\langle P \rangle$. The rest of this section details this analogy (see also Section 2.3.3).

The following definition of a discrete convex hull is given and is to be related to the definitions of convexity above.

Definition 2.38: L- and T-convex hulls [132]

Given a set of discrete points P, the L-convex hull (respectively T-convex hull) of P is the intersection of all discrete L-convex (respectively T-convex) sets that contain P.

Since $\langle P \rangle$ is T-convex and minimal, the T-convex hull is characterised by $\langle P \rangle$. For the characterisation of the L-convex hull of P, we refer to the notation introduced in [132].

Definition 2.39: Notation for the L-convex hull [132]

Given a set of discrete points P,

$$L(P) = \bigcup_{p,q \in P} \langle p, q \rangle$$

and for any integer $m \geq 1$, $L^m(P)$ is defined recursively by

$$L^m(P) = \begin{cases} L(P) & \text{if } m = 1 \\ L(L^{m-1}(P)) & \text{if } m > 1 \end{cases}$$

Remark 2.40:

If $|P|$ is finite, there exists m^* such that $L^m(P) = L^{m^*}(P)$ for any integer $m \geq m^*$.

Using this notation, it is clear that the L-convex hull $\langle P \rangle_{\mathrm{L}}$ of a set of discrete points P can be characterised as follows.

Proposition 2.41: [132]

Given a set of discrete points P, its L-convex hull $\langle P \rangle_{\mathrm{L}}$ is characterised by

$$\langle P \rangle_{\mathrm{L}} = \bigcup_{m \geq 1} L^m(P)$$

It is easily seen that $\langle P \rangle_{\mathrm{L}} \subseteq \langle P \rangle$. The equality is obtained under additional conditions given by Proposition 2.42. Note that these conditions (concerning connectivity) are consistent with that necessary for the equivalence of discrete convexity definitions (e.g. see Figure 2.22).

Proposition 2.42: [132]

Given a set of discrete points P,

(i) If $\langle P\rangle_{\mathrm{L}}$ is 8-connected, then $\langle P\rangle_{\mathrm{L}} = \langle P\rangle$.

(ii) If P is 8-connected, then $\langle P\rangle_{\mathrm{L}} = \langle P\rangle$ and $\langle P\rangle$ is 8-connected.

Proof:

(i) follows directly from Proposition 2.33(ii).

(ii): If P is 8-connected, there exists an 8-connected component $Q \subseteq \langle P\rangle_{\mathrm{L}}$ such that $P \subseteq Q$. By equivalence of 8- and L-convexity under the 8-connectivity condition (see Figure 2.22), Q is L-convex. From Proposition 2.33(ii), it follows that Q is T-convex. Therefore, $\langle P\rangle \subseteq \langle Q\rangle = Q \subseteq \langle P\rangle_{\mathrm{L}}$. Since $\langle P\rangle_{\mathrm{L}} \subseteq \langle P\rangle$, clearly $\langle P\rangle_{\mathrm{L}} = \langle P\rangle = Q$. Moreover, $\langle P\rangle$ is 8-connected. Therefore, Proposition 2.42 holds. □

Figure 2.24 repeats the example shown in Figure 2.21 and shows the difference between L- and T-convex hulls. In this example, $P = \{p, q, r\}$. Figure 2.24(A) shows that the L-convex hull of P is composed of P itself. Therefore, $\langle P\rangle_{\mathrm{L}} = P$. On the other hand, Figure 2.24(B) shows that the T-convex hull of P is $\langle P\rangle = \{p, q, r, s\}$. Therefore, $\langle P\rangle = P \cup \{s\} = \langle P\rangle_{\mathrm{L}} \cup \{s\}$.

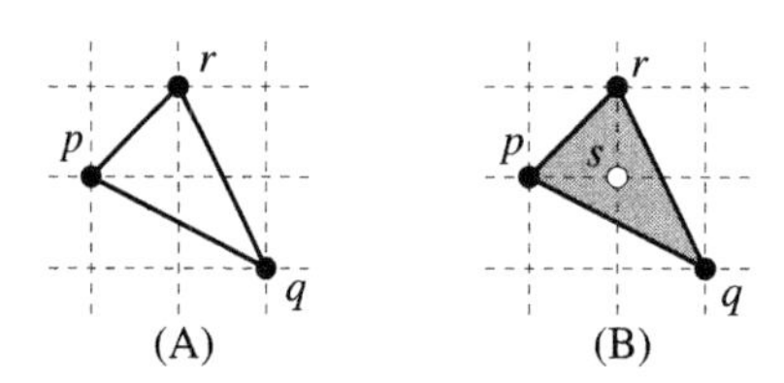

Figure 2.24 Relation between (A) L-convex and (B) T-convex hulls

Considering the variety of definitions for discrete convexity given above, it is of interest to see whether corresponding definitions for the discrete convex hull (based on Definition 2.37) yield the same discrete set. This is mostly the case, as pointed out in [132].

Proposition 2.43: [132]

Let P be a set of discrete points and $k = 4$ or 8. The intersection of all k-convex sets that contain P is $\langle P\rangle$. Equivalently, the k-convex hull of P is $\langle P\rangle$.

In summary, definitions for the discrete convex hull of a set of discrete points P mostly lead to the characterisation of $\langle P\rangle$, the set of discrete points contained in $[P]$, the continuous convex hull of P. This is true when the definition of the discrete convex hull is based on cellular, T- or k-convexity. The stronger case of L-convexity proves also to be equivalent under 8-connectivity. Therefore, in general, the discrete convex hull of a set of discrete points P refers to $\langle P \rangle$.

An important study of computational geometry is to derive algorithms that compute the discrete convex hull of P (e.g. see [20, 65, 120, 160, 177] and Section 6.2.2). As a concluding example, Figure 2.25 displays an example for the construction of the discrete convex hull of a set of discrete points P. In Figure 2.25, points in P are displayed as black discs (•).

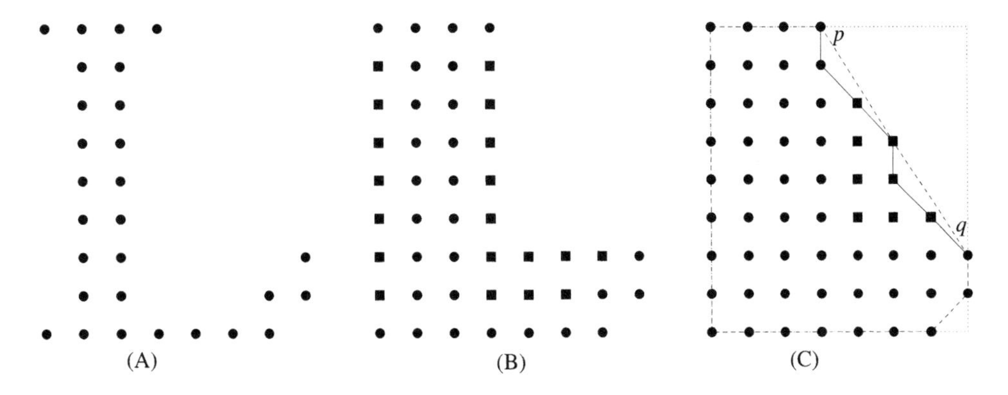

Figure 2.25 Discrete convex hull of a discrete non-convex set

Clearly, in this example, P is not discrete convex. The discrete convex hull can be constructed in two steps. Firstly, for every column i in P, extremal points (i, y^i_{min}) and (i, y^i_{max}) are defined as the lower-most and upper-most points of P in column i. Similarly, extremal points (x^j_{min}, j) and (x^j_{max}, j) are the left-most and right-most points of P in the line j, respectively. P is then updated by filling each line and column between their respective extremal points. The resulting set P' is then said to be (x, y)-convex. Clearly, $P \subseteq P'$ and P' does not contain any hole. Figure 2.25(B) shows the set P', where points that have been added to points in P are represented as black squares. Finally, the minimum enclosing vertical rectangle of P' is characterised. In Figure 2.25(C), this rectangle is shown as a dotted line. Extremal points that belong to this rectangle may define straight segments that are not included in P' (e.g. points p and q in Figure 2.25(C)). By operations on the coordinates of p and q, points forming the digital straight segment between p and q can be added to P'. Repeating this operation for any combination of extremal points defines the discrete convex hull of P' and therefore $\langle P \rangle$.

2.3.3 Compatibility

Compatibility of the above definitions of discrete convexity with that of continuous convexity is investigated here. This is achieved via alternative definitions of discrete convexity which prove again to be equivalent to the ones presented above (we purposely left them aside in the previous section since they refer to a digitisation scheme and are not geometrical characterisations of discrete convexity).

A first trivial definition of discrete convexity related to a digitisation scheme is given below.

Definition 2.44: Discrete convexity [159]

A set of discrete points P is convex if and only if it is the digitisation of a continuous convex set.

Definition 2.44 relies on the definition of a digitisation scheme. The square-box quantisation presented in Section 2.2.2 and studied in Chapter 4 is used for this purpose. Figures 2.26(A) and (B) illustrate the digitisation of two continuous convex and non-convex sets, respectively.

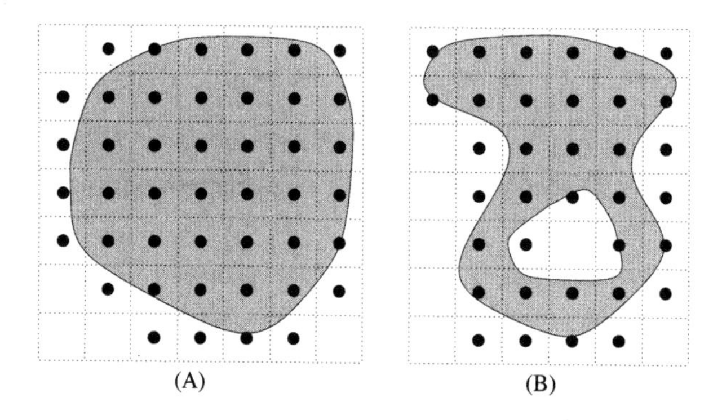

Figure 2.26 Square-box quantisation of continuous (A) convex and (B) non-convex sets

In this context, one can prove that the square-box quantisation of a continuous convex set is T-convex. Conversely, given a T-convex set P, let S be the union of all digitisation boxes associated with discrete points in P and let ∂S be its border. It has been proved that, if a T-convex set P does not contain any discrete point p associated with a digitisation box which has more than two sides in common with ∂S, then P is the square-box quantisation of at least one continuous convex set S [65].

Figure 2.27(A) shows an instance where P is T-convex and where there does not exists continuous convex set whose square-box quantisation yields P. Points in P are displayed as black discs (•). The discrete point $p \in P$ is associated with a digitisation box that has three sides in common with ∂S (displayed as a thick dashed contour). Figure 2.27(B) displays an example where such a point $p \in P$ exists and where there exists a continuous convex set whose square-box quantisation yields P. This continuous set S is shown as a shaded area.

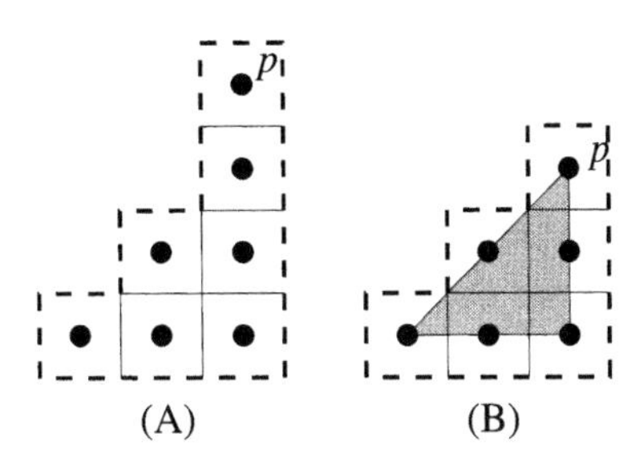

Figure 2.27 Relation between continuous and T-convexity

The previous definition creates a relationship between continuous and discrete convexity. However, no strict condition is given for a given set of discrete points to be the digitisation of a continuous convex set. The last part of this section addresses this problem. The definition of (discrete) ε-convexity is given and depends upon the sampling step used for the digitisation scheme [20] (see Definition 4.3).

Definition 2.45: ε-Convexity [20]

Given a sampling step $h > 0$ and a real value $\varepsilon \in [h, \frac{h}{2}[$, a set of discrete points P is ε-convex if and only if for any discrete points p_i and p_j in P, there exists a discrete point q in P such that, for any real point α on the continuous segment $[p_i, p_j]$, $d_8(q, \alpha) \leq \varepsilon$.

Remark 2.46:

Note the similarity between this definition and that of the chord property (i.e. Proposition 2.6 where $\varepsilon = 1$).

It is proved that the ε-convex hull of a set of discrete points P is exactly equal to $\langle P \rangle$. In this sense, ε-convexity is equivalent to T-convexity. ε-Convexity could therefore be included in Figure 2.22 as a definition of discrete convexity equivalent to L-, k- or cellular convexity via appropriate additional conditions. The main result concerning the study of compatibility between discrete and continuous convexity can now be formulated as follows.

Proposition 2.47: [20]

Given a continuous set S, if for every sampling step $h > 0$, the square-box quantisation of S is discrete convex then S is convex in the continuous space.

This result is now equivalent to that presented in Definition 2.44. In other words, there exists an equivalence relationship between continuous and discrete convexity. Continuous convexity is seen as the "limit case" of discrete convexity (i.e. when the sampling step tends to zero).

In summary, this section reviewed a collection of definitions for discrete convexity and showed their equivalence via additional connectivity conditions. The discrete convex hull has also been defined and shown to be intimately related to its continuous counterpart. Finally, compatibility between both continuous and discrete convexity definitions has been proved via the definition of ε-convexity and the use of the square-box quantisation used as the underlying digitisation scheme.

2.4 Discrete curvature

In Section 2.2, the characterisation of a discrete straight line was presented. Another important characteristic for curve-like objects is their curvature. This, in turn, allows for the characterisation of circles which themselves characterise rotations. In this section, these concepts are first briefly reviewed from a geometrical viewpoint in the continuous space $\mathbb{R}^2$ and mapped onto the discrete space $\mathbb{Z}^2$ represented by the square lattice.

2.4.1 A note on continuous curvature

In the continuous space $\mathbb{R}^2$, the curvature of a curve representing the parametric function $f(t) = (x(t), y(t))$ at $t \in \mathbb{R}$ is given by

$$\rho_f(t) = \frac{|y''(t)x'(t) - x''(t)y'(t)|}{(x'(t)^2 + y'(t)^2)^{\frac{3}{2}}}$$

Geometrically, the curvature $\rho_f(t)$ represents the inverse of the radius of the best tangential circle which approximates the curve represented by f at the point $f(t) = (x(t), y(t))$ (e.g. see Figure 2.28). The centre of this circle is located at $c(t)$ such that $\overrightarrow{f(t)c(t)} = \frac{1}{\rho_f(t)}\vec{n}(t)$, where $\vec{n}(t)$ is the unit vector normal to f at the point $f(t)$.

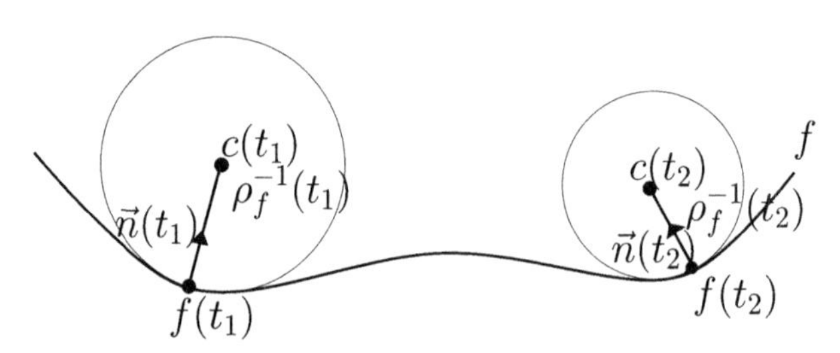

Figure 2.28 Continuous curvature

In this case, the curvature of a circle is constant at any point and its value is the radius of the circle itself. The special case of a straight line leads to a zero-curvature and, hence, an infinite radius at any point of the line.

2.4.2 Discrete curvature

In the discrete space, curvature commonly refers to changes of direction within the discrete object under study. Hence, discrete curvature only makes sense when it is possible to define an order for a sequence of discrete points. Therefore, the study of discrete curvature will be restricted to digital arcs and curves. The uniqueness of the connectivity relationships between points in a digital arc or a digital curve readily defines an order on these points (e.g. see chain-code in Section 2.2.1). Changes of direction will be measured via angles made between specific continuous segments defined by points on the discrete object in question.

Definition 2.48: Discrete curvature [22, 139]

Given a set of discrete points $P = \{p_i\}_{i=0,\ldots,n}$ that defines a digital arc or a digital curve, $\rho_k(p_i)$, the k-order curvature of P at point $p_i \in P$ is given by $\rho_k(p_i) = 1 - |\cos\theta_k^i|$ where $\theta_k^i = \text{angle}(p_{i-k}, p_i, p_{i+k})$ is the angle made between the continuous segments $[p_{i-k}, p_i]$ and $[p_i, p_{i+k}]$ and $k \in \{i, \ldots, n-i\}$.

Remark 2.49:

For a digital closed curve, restrictions on the order k can be ignored if calculating point indices modulo $n+1$ (i.e. $p_{n+1} = p_0$, $p_{n+2} = p_1$ and so on).

By definition, the discrete curvature value increases as the digital arc deviates from a straight segment. It is therefore expected that a digital straight segment will be associated with a zero-value curvature at every point.

The integer order k is introduced to make the curvature less sensitive to local variations in the directions followed by the digital arc or digital curve. A discrete curvature of high order will reflect more accurately the global curvature of the continuous object approximated by the set of discrete points. Figure 2.29 illustrates the process of calculating $\rho_3(p_{10})$, the 3-order discrete curvature of the digital arc $P_{pq} = \{p_i\}_{i=0,\ldots,17}$ at point p_{10}.

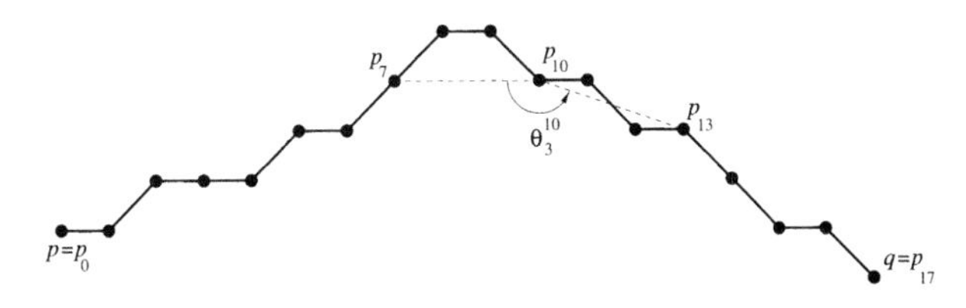

Figure 2.29 Discrete curvature calculation. Case of $\rho_3(p_{10})$

Figure 2.30 displays the resulting curvature values obtained for successive instances of the order k ($k = 1, \ldots, 6$) when using the digital arc presented in Figure 2.29. Clearly, the 1-order curvature is not an accurate representation for discrete curvature since it only highlights local variations. This example also shows that as the order k increases, the curvature tends to illustrate the global behaviour of the digital arc. A peak of discrete curvature is detected at point p_8 or p_9. Clearly, these discrete points correspond to the location where a dramatic change of global direction occurs in the digital arc.

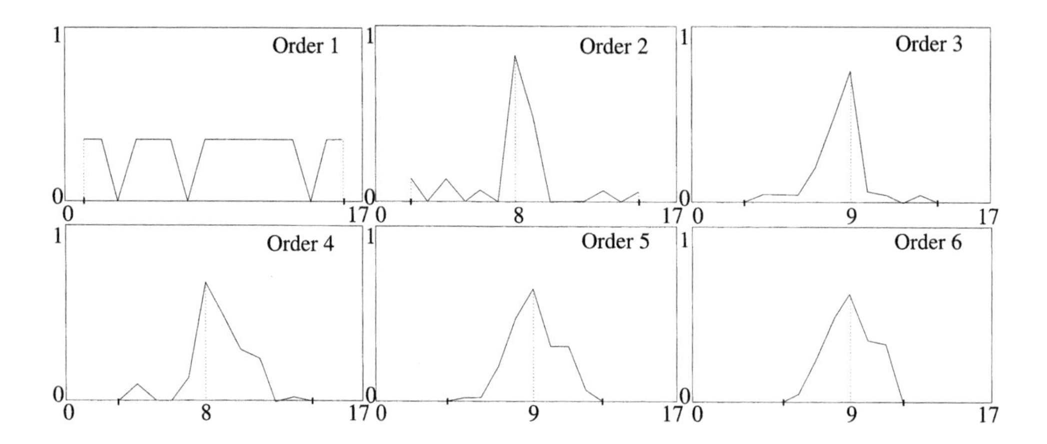

Figure 2.30 Curvature of different orders

In the case of a digital straight segment, points are repeated periodically (see Propositions 2.4 and 2.5). Therefore, if the order k of the calculated curvature matches this period, the curvature will be computed using $\theta_k^i = \pi$ for all i and hence, $\rho_k(p_i) = 0$, for all i. Otherwise, the same calculation will be repeated along the digital straight segment and the discrete curvature will vary in the same periodic fashion as in the digital straight segment. This is illustrated by Figure 2.31. The digital straight segment P_{pq} in question is shown in Figure 2.31(A). Its period is clearly three moves long (i.e. involving a subsequence of four points).

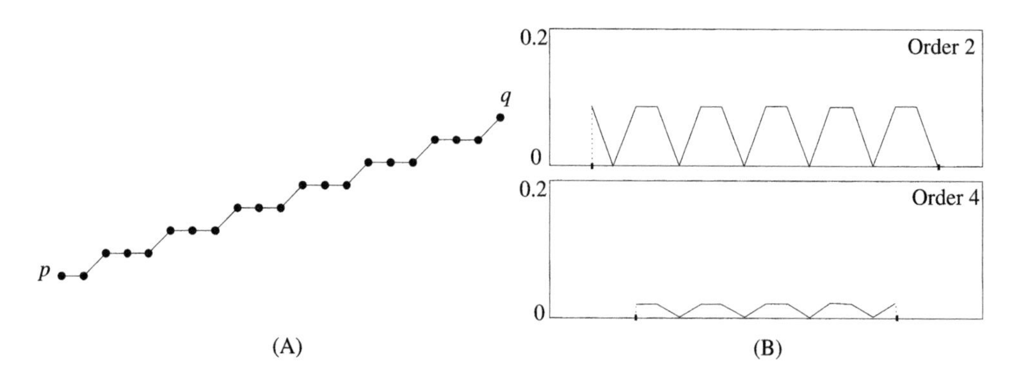

Figure 2.31 Curvature of a digital straight segment

The digital arc P_{pq} shows a three-moves-long period. As expected, the 3-order curvature (and, more generally, the $3k$-curvature) of P_{pq} is null at every point of P_{pq}. Moreover, this period is apparent in Figure 2.31(B) which shows the 2- and 4-order curvatures of P_{pq}. Note that the average value of the curvature tends towards zero as the order increases.

2.4.3 Discrete circles

Unlike their continuous counterparts, discrete circles do not refer to any corresponding discrete distance. Discrete circles are not defined as set of points at a given discrete distance value from a given centre. Rather, they are defined as digitisation of continuous circles. In this respect, care must be taken when relating discrete discs defined by Definition 1.20 to discrete circles. Definitions and characterisations of discrete circles resulting from grid-intersect and square-box quantisation schemes are given here.

Definition 2.50: Discrete circle (using grid-intersect quantisation) [104]

Given a set of discrete points P and a discrete point c, there exists a continuous circle centred at c whose grid-intersect quantisation equals P if and only if the following conditions are satisfied.

(i) For each octant, the subset of P which lies in this octant defines an 8-digital arc.

(ii) Let $r = (x_r, y_r)$ and $s = (x_s, y_s)$ be two discrete points in P such that

$$d_{\mathrm{E}}(c,r) = \max_{p_i \in P}\{d_{\mathrm{E}}(c,p_i)\} \quad \text{and} \quad d_{\mathrm{E}}(c,s) = \min_{p_i \in P}\{d_{\mathrm{E}}(c,p_i)\}$$

If r' and s' are two real points defined by

$$\begin{cases} r' = (x_r - \frac{1}{2}, y_r) & s' = (x_s + \frac{1}{2}, y_s) & \text{if } x_r \geq 0 \text{ and } x_r \geq |y_r| \\ r' = (x_r, y_r - \frac{1}{2}) & s' = (x_s, y_s + \frac{1}{2}) & \text{if } y_r \geq 0 \text{ and } y_r \geq |x_r| \\ r' = (x_r + \frac{1}{2}, y_r) & s' = (x_s - \frac{1}{2}, y_s) & \text{if } x_r \leq 0 \text{ and } |x_r| \geq |y_r| \\ r' = (x_r, y_r + \frac{1}{2}) & s' = (x_s, y_s - \frac{1}{2}) & \text{if } y_r \leq 0 \text{ and } |y_r| \geq |x_r| \end{cases}$$

Then $d_{\mathrm{E}}(c, r') < d_{\mathrm{E}}(c, s')$.

Figure 2.32 illustrates an example of discrete circle P defined via the grid-intersect quantisation.

Points in P are displayed by black discs (•). Points r and s are shown with their respective shifted images r' and s'. Dashed circular arcs illustrate the fact that $d_{\mathrm{E}}(c,r') < d_{\mathrm{E}}(c,s')$. A possible continuous circle centred at c whose grid-intersect quantisation yields P is also shown as a continuous thin circle.

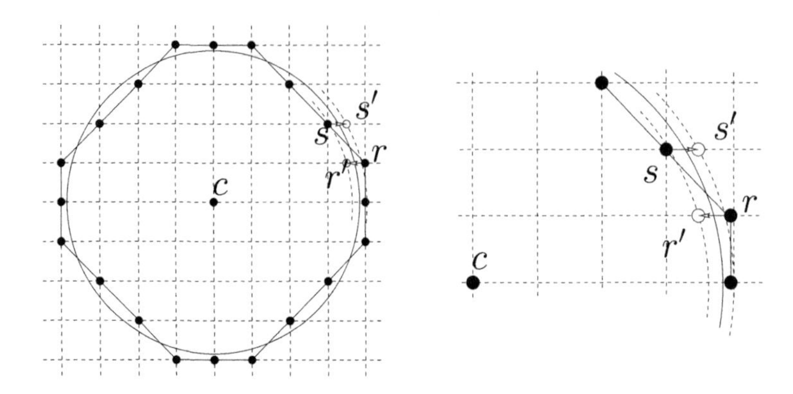

Figure 2.32 Discrete circle defined via the grid-intersect quantisation

Definition 2.51: Discrete circle (using square-box quantisation) [104]

Given a set of discrete points P and a discrete point c, there exists a continuous circle centred at c whose square-box quantisation equals P if and only if the following conditions are satisfied.

(i) For each octant, the subset of P which lies in this octant defines an 4-digital arc.

(ii) Let $p_i = (x_i, y_i) \in P$. Then r_i and s_i are two real points defined by

$$\begin{cases} r_i = (x_i + \frac{1}{2}, y_i + \frac{1}{2}) & s_i = (x_i - \frac{1}{2}, y_i - \frac{1}{2}) & \text{if } x_i \geq 0 \text{ and } x_i \geq |y_i| \\ r_i = (x_i - \frac{1}{2}, y_i + \frac{1}{2}) & s_i = (x_i + \frac{1}{2}, y_i - \frac{1}{2}) & \text{if } y_i \geq 0 \text{ and } y_i \geq |x_i| \\ r_i = (x_i - \frac{1}{2}, y_i - \frac{1}{2}) & s_i = (x_i + \frac{1}{2}, y_i + \frac{1}{2}) & \text{if } x_i \leq 0 \text{ and } |x_i| \geq |y_i| \\ r_i = (x_i + \frac{1}{2}, y_i - \frac{1}{2}) & s_i = (x_i - \frac{1}{2}, y_i + \frac{1}{2}) & \text{if } y_i \leq 0 \text{ and } |y_i| \geq |x_i| \end{cases}$$

Then $\min\{d_E(c, r_i) \; ; \; p_i \in P\} > \max\{d_E(c, s_i) \; ; \; p_i \in P\}$.

Figure 2.33 illustrates an example of a discrete circle P obtained via the square-box quantisation.

Points in P are displayed as black discs (•). Points r and s are shown and represent the points r_i and s_i for which the minimum and maximum defined in Definition 2.51 are obtained, respectively. Dashed circular arcs illustrate the fact that $d_E(c, r) > d_E(c, s)$. A possible continuous circle centred at c whose square-box quantisation yields P is also shown as a continuous thin circle.

Remark 2.52:

By definition of r and s, any continuous circle centred at c that has a radius $\rho > 0$ such that $d_E(c, s) \leq \rho \leq d_E(c, r)$ is mapped onto P by the square-box quantisation.

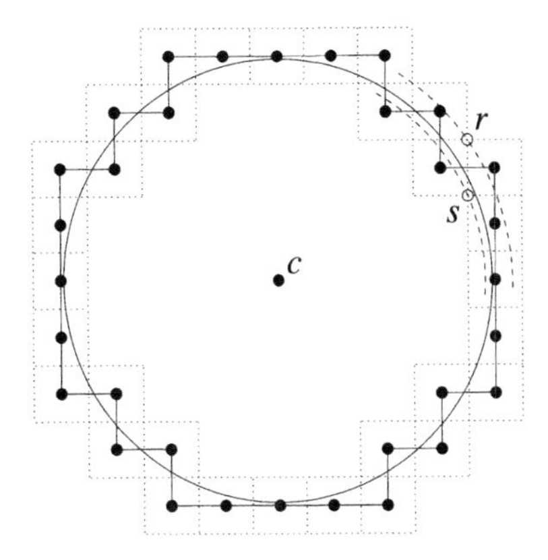

Figure 2.33 Discrete circle defined via the square-box quantisation

In the continuous space, circles show a constant non-zero curvature at any of their points. The definition of discrete curvature allows for a similar characterisation of discrete circles. In the following example, the discrete curvature of a discrete circle is calculated for different values of the order k. Figure 2.34(A) shows the upper-half of the discrete circle centred at c used for the calculation (the lower-half is obtained by symmetry). Figure 2.34(B) displays the values of the 3- and 5-order curvatures of this discrete circle.

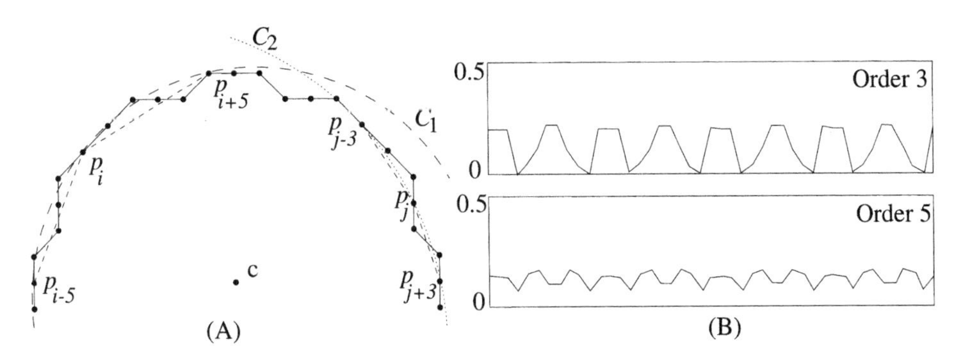

Figure 2.34 Curvature of a discrete disc. (A) Digital circle and calculation of the 5-order curvature. Approximation circles are also shown dotted and dashed. (B) Values of the 3- and 5-order curvatures

Both 3- and 5-order curvatures show a periodic behaviour. This is explained by the symmetry of the discrete circle. The same digital arc is repeated in each octant through rotations or symmetry. The 5-order curvature shows an oscillation around a value of 0.15.

In this case, $k = 5$ is the optimal value to calculate the discrete curvature of this discrete circle since the angle $\theta_5^i = \text{angle}(p_{i-5}, p_i, p_{i+5})$ accurately approximates the underlying continuous circle (see Figure 2.34(A)). In other words, the continuous circles defined by p_{i-5}, p_i and p_{i+5} are similar in radius to the underlying continuous circle whose digitisation is P (see dashed circle C_1 in Figure 2.34(A)). By contrast, the 3-curvature calculated via θ_3^j = angle(p_{j-3}, p_j, p_{j+3}) is sensitive to local variations within the discrete circle. Moreover, p_{j-3}, p_j and p_{j+3} do not define an accurate approximation for the discrete circle in question (see dotted circle C_2 in Figure 2.34(A)). On the other hand, an excessive value for the order of the curvature would not create a relevant calculation scheme. A trade-off in the choice of the order is therefore necessary when calculating curvature of digital arcs.

Discrete circles are defined as digitisation of continuous circles. Therefore, they are shown to have geometrical characteristics that are equivalent to those of their continuous counterparts. Discrete characterisations exist and are typically used for deriving algorithms which can recognise whether a set of discrete points represents a discrete circle, or otherwise (see, for example, [68, 104, 105]). The converse also exists. Algorithms that define iteratively discrete circles have been suggested (e.g. see [7, 16, 17]). Improvements of such algorithms have been proposed (e.g. see [81, 172]). We highlight [81], where the characterisation of "gaps" is introduced. Gaps are discrete points that do not belong to any discrete circle generated by the algorithms cited above. Discrete rotations are similarly defined as discrete mappings of continuous rotations. A purely discrete rotation is still to be defined.

As an application of the above results, it is shown in Chapter 6 how discrete curvature can be used for characterising a shape via its angular points.

2.5 Parallelism and orthogonality

In the discrete space, properties such as parallelism and orthogonality apply to digital straight segments. Given two digital straight segments P_{pq} and $P_{p'q'}$, these properties are characterised via the definition of continuous segments whose digitisation equals P_{pq} and $P_{p'q'}$, respectively. Such continuous segments are referred to as preimages of P_{pq} and $P_{p'q'}$ respectively (see Definition 4.1 for more details).

Typically, for a given digital straight segment P_{pq}, the domain of P_{pq} refers to the set of continuous segments whose digitisation equals P_{pq}. A calculation method for the domain of a digital straight segment is given in [37] and a simpler proof for it is detailed in [79]. In Section 4.1, the general concept of the domain of a set of discrete points is detailed. In this section, we briefly introduce this concept with the aim of using it in the characterisation of discrete parallelism and orthogonality.

Let us assume that the grid-intersect quantisation (GIQ) is used as the digitisation scheme to characterise P_{pq}. Figure 2.35 displays the GIQ-domain of

P_{pq}. This set is first illustrated in Figure 2.35(A) as directly characterised by Definition 4.1.

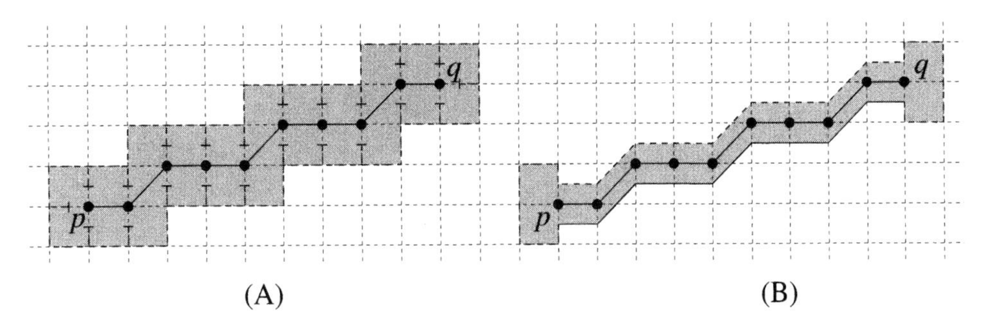

Figure 2.35 GIQ-domain of a digital straight segment

Note the similarity between the GIQ-domain of P_{pq} and the visibility polygon defined by the chord and compact chord properties on P_{pq} (see also Remark 4.15). Since in this case, preimages of P_{pq} are to be continuous straight segments, a trivial reduction of the GIQ-domain of P_{pq} results in the area shown in Figure 2.35(B). Alternative continuous and dashed borders of the shaded area highlight the fact that this set is semi-open (see Section 4.1.3 for more details). Further minimisation of this set is detailed in [37].

Let $\sigma_{\min}$ and $\sigma_{\max}$ be the minimum and maximum of slopes of continuous segments $[\alpha, \beta]$ which are preimages of P_{pq}. Both segments $[\alpha_{\min}, \beta_{\min}]$ and $[\alpha_{\max}, \beta_{\max}]$ which correspond to slopes $\sigma_{\min}$ and $\sigma_{\max}$, respectively, are used to characterise parallelism and orthogonality [79, 105].

Proposition 2.53: Digital parallelism and orthogonality [79]

Given two digital straight segments P_{pq} and $P_{p'q'}$, they are parallel (respectively orthogonal) if and only if there exists two parallel (respectively orthogonal) continuous segments included in the GIQ-domains of P_{pq} and $P_{p'q'}$, respectively.

Based on digital parallelism and orthogonality, it is possible to define digital squares or rectangles and to derive algorithms which recognise such figures (see, for example, [79, 105, 106]). As a further remark, it should be noted that the intersection between digital straight segments is also characterised via their continuous counterparts. It can be the case that, although two continuous segments intersect each other, the intersection of their respective digitisation sets are empty [22] (see Figure 2.36). For more details, the interested reader is referred to [79, 105, 106].

In conclusion, discrete geometry allows for the characterisation of well-known geometrical properties on discrete objects. Discrete properties are based on discrete points with the aim of avoiding reference to continuous objects. However, such an ideal characterisation is not always possible and some concepts such

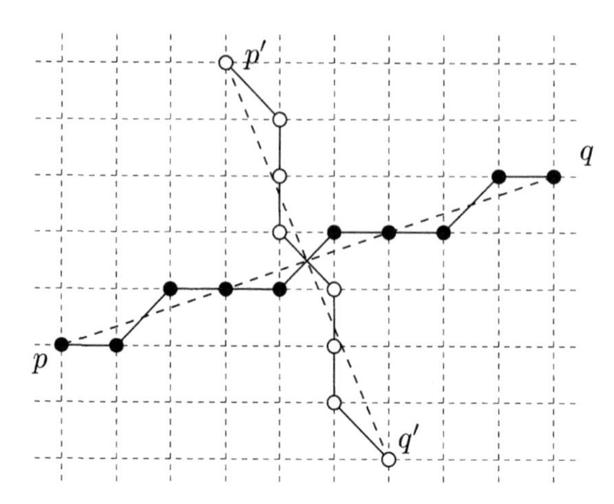

Figure 2.36 Intersection of two digital straight segments

as rotations, orthogonality or parallelism are defined via the continuous counterparts of the discrete objects in question.

The problem of digitisation will be studied further when addressing the topic of image acquisition and storage in Chapter 4. Morphological characterisation of a discrete set of points (i.e. shape analysis) will be developed in Chapter 6.

Chapter 3

ALGORITHMIC GRAPH THEORY

Graph theory is an important mathematical approach which can be used for mapping complex problems onto simple representations and models. It is based on a robust mathematical background which allows for the definition of optimal solution techniques for such problems. Efficient algorithms can thus be derived which can solve a particular problem based on its graph representation.

Graph theory is an area of research by itself and finds applications in many fields such as operational research and theoretical computer science. It also finds applications in image processing, where the discrete nature of image representations makes its use consistent. In fact, it is often the case that graph theory is implicitly used when developing a solution to a particular problem related to image processing. The aim here is to relate image processing concepts to algorithmic graph theory in order to take advantage of this approach for further developments.

In this chapter, we first present in Section 3.1 terminology and definitions used in graph theory. We summarise well-known algorithms which exist for the solution of problems defined in the context of graph theory. Typically, classes of equivalent problems are defined and solutions are presented for abstract representations of each class. Section 3.2 presents algorithms for the solution of two classic optimisation problems. More precisely, a number of algorithms are presented for the shortest path problem and the minimum weighted spanning tree problem. Finally, Section 3.3 relates these results to digital image processing concepts presented in Chapters 1 and 2.

3.1 Definitions

Major definitions used in algorithmic graph theory are reviewed here. The concept of a graph is introduced via the definition of vertices and arcs. These concepts are then extended to that of paths and trees on a graph. Further details on the definitions and properties summarised here can be found in most books on discrete and combinatorial mathematics (see, for example, [25] and [49]).

3.1.1 Graphs

Graphs are used to give efficient representations to problems that involve items and inter-relations between them. In graph theory, items and relations are typically represented by vertices and arcs, respectively. Such a representation is called a graph.

Definition 3.1: Graph

(i) *A graph $G = (V, A)$ is a set of vertices V with their inter-relationships given by the set of arcs A. If an orientation is associated with any arc, the graph is said to be directed, otherwise G is an undirected graph.*

(ii) *An arc connecting two vertices u and v in V is denoted as (u, v). If (u, v) is a directed arc, then, by convention, it is oriented from vertex u to vertex v. If (u, v) is an undirected arc, then $(u, v) = (v, u)$*

(iii) *If (u, v) is an arc in A, vertex v is called a successor of vertex u and vertex u is called a predecessor of vertex v. Moreover, u and v are said to be adjacent in G. The set of vertices adjacent to a vertex $u \in V$ (i.e. the set of successors of u) forms the forward star of vertex u. Similarly, two arcs are said to be adjacent if they have at least one vertex in common.*

(iv) *N_G and M_G (or N and M where no confusion arises) are defined as the number of vertices and arcs in the graph $G = (V, A)$, respectively (i.e. $N_G = |V|$ and $M_G = |A|$). If $M_G = \mathcal{O}(N_G)$ then the graph G is said to be sparse. If for any vertex $u \in V$, there exists an arc $(u, v) \in A$ connecting u to every other vertex $v \in V$, the graph $G = (V, A)$ is said to be complete.*

Remark 3.2:

To each directed graph $G = (V, A)$ corresponds an undirected graph $\overline{G} = (V, \overline{A})$ where $\overline{A}$ is the set of arcs in A without their orientations.

Example: Graph

Consider the set of four locations $\{v_1, v_2, v_3, v_4\}$. Consider the directed adjacency relationship where a location is visible from another one, or otherwise. A summary of these relations is given in Table 3.1.

Location	Visible location
v_1	v_2,v_3,v_4
v_2	v_3
v_3	none
v_4	v_1,v_3

Table 3.1 Visibility relationships between locations

The graph representation of this data is given in Figure 3.1(A). The set V of vertices in this graph is the set of locations, $V = \{v_1, v_2, v_3, v_4\}$. Therefore, $N_G = 4$. Vertices are represented by black discs ($\bullet$) in Figure 3.1. The set of arcs A in the graph corresponds to the set of visibility relations. An arc $(u, v) \in A$ exists in the graph $G = (V, A)$ between two vertices $u \in V$ and $v \in V$ if and only if the location corresponding to vertex v is visible from the location represented by vertex u. The orientation of arcs is represented by arrows in Figure 3.1(A). In this example, the set of arcs is $A = \{(v_1, v_2), (v_1, v_3), (v_1, v_4), (v_2, v_3), (v_4, v_1), (v_4, v_3)\}$. Hence, $M_G = 6$.

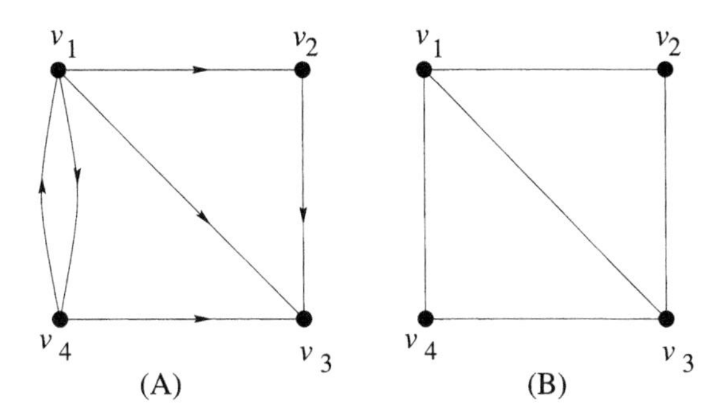

Figure 3.1 Graph. (A) Directed. (B) Undirected

Suppose now that the locations are visible in both directions. The resulting graph $\overline{G} = (V, \overline{A})$ is therefore undirected since each arc can be traversed in either direction. In this case, the graph representation of this data is as given in Figure 3.1(B). The resulting set of undirected arcs is

$$\overline{A} = \{(v_1, v_2), (v_1, v_3), (v_1, v_4), (v_2, v_3), (v_3, v_4)\}$$

and $M_{\overline{G}} = 5$. Clearly, $N_{\overline{G}} = N_G = 4$ since both G and $\overline{G}$ are based on the same set of vertices V.

Graphs can be represented in different ways. For example, Table (3.1) gives the representation of G by a list of successors for each vertex. Another important graph representation maps the graph onto an adjacency matrix.

Definition 3.3: Adjacency matrix

A graph $G = (V, A)$ where $V = \{u_0, u_1, \ldots, u_{N-1}\}$ can be represented by its adjacency matrix $\mathcal{M}(G) = (m_{ij})_{i=0,\ldots,N-1, j=0,\ldots,N-1}$ defined by

$$m_{ij} = \begin{cases} 1 \text{ if } (u_i, u_j) \in A \\ 0 \text{ if } (u_i, u_j) \notin A \end{cases}$$

Clearly, the adjacency matrix of an undirected graph is symmetric (i.e. $m_{ij} = m_{ji}\ \forall i, j$).

Example: Adjacency matrix

Continuing with the earlier example of a graph, the adjacency matrix of the directed graph G shown in Figure 3.1(A) is

$$\mathcal{M}(G) = \begin{pmatrix} 0 & 1 & 1 & 1 \\ 0 & 0 & 1 & 0 \\ 0 & 0 & 0 & 0 \\ 1 & 0 & 1 & 0 \end{pmatrix}$$

and the adjacency matrix of the undirected graph $\overline{G}$ shown in Figure 3.1(B) is

$$\mathcal{M}(\overline{G}) = \begin{pmatrix} 0 & 1 & 1 & 1 \\ 1 & 0 & 1 & 0 \\ 1 & 1 & 0 & 1 \\ 1 & 0 & 1 & 0 \end{pmatrix}$$

Different results and properties are obtained by applying linear algebraic operations on the graph using its adjacency matrix (e.g. see [29]).

Proposition 3.4:

Given a directed graph $G = (V, A)$ and its adjacency matrix $\mathcal{M}(G) = (m_{ij})_{i,j}$, the corresponding undirected graph $\overline{G}$ is associated with the adjacency matrix $\mathcal{M}(\overline{G}) = (\overline{m}_{ij})_{i,j}$, where $\overline{m}_{ij} = \max\{m_{ij}, m_{ji}\}$.

In a graph, arcs are generally associated with values representing the cost of the relationships represented by the arcs. Depending on the application considered, an arc cost, arc weight or arc length can be used to label this value.

Definition 3.5: Arc weight and arc length

The weight $w(u, v)$ of the arc (u, v) is a real value associated to it. The arc length $l(u, v)$ is a particular case of weight for the arc (u, v).

Remark 3.6:

(i) The adjacency matrix of the graph $G = (V, A)$ can also be used to store arc weights when redefining m_{ij} as follows (compare with Definition 3.3):

$$m_{ij} = \begin{cases} w(u_i, u_j) & \text{if } (u_i, u_j) \in A \\ 0 & \text{if } (u_i, u_j) \notin A \end{cases}$$

(ii) The notation w_{ij} (respectively l_{ij}) will be equivalently used for the weight $w(u_i, u_j)$ (respectively the length $l(u_i, u_j)$) of the arc (u_i, u_j).

Example: Arc weight and arc length

In the earlier example of a graph, a possible weight for an arc (v_i, v_j) is the minimal distance from a point of the segment $[v_i, v_j]$ to the border of the visibility polygon. The arc length is readily defined by the distance between two locations. Consider the values of weight and length given by Table 3.2.

Arc	Weight	Length
v_1 to v_2	2	3
v_1 to v_3	4	5
v_1 to v_4	3	5
v_2 to v_3	6	7
v_4 to v_1	3.5	6
v_4 to v_3	5	5

Table 3.2 Relationships between locations

Then, the adjacency matrix of $G = (V, A)$ becomes

$$\mathcal{M}^l(G) = \begin{pmatrix} 0 & 2 & 4 & 3 \\ 0 & 0 & 6 & 0 \\ 0 & 0 & 0 & 0 \\ 3.5 & 0 & 0 & 5 \end{pmatrix} \qquad \mathcal{M}^w(G) = \begin{pmatrix} 0 & 3 & 5 & 5 \\ 0 & 0 & 7 & 0 \\ 0 & 0 & 0 & 0 \\ 6 & 0 & 0 & 5 \end{pmatrix}$$

when using length and weight data, respectively.

3.1.2 Paths and trees

The representation given by the graph $G = (V, A)$ can be readily used to define a sequence of vertices. Such a sequence generally characterises a solution to a specific problem and is referred to as a path in the graph.

Definition 3.7: Path

(i) *A path P_{uv} on the graph $G = (V, A)$ from vertex $u \in V$ to vertex $v \in V$ is a set of vertices $P_{uv} = \{u_0, u_1, \ldots, u_n\} \subseteq V$ such that $u_0 = u$, $u_n = v$ and $(u_i, u_{i+1}) \in A$ for all $i \in \{0, \ldots, n-1\}$. In the specific instance where each vertex is visited exactly once, the path is called a simple path (i.e. $u_i \neq u_j$ for any pair of distinct vertices $u_i, u_j \in P_{uv}$).*

(ii) *If all arcs in the path are undirected, this undirected path is equivalently called a chain.*

(iii) *The value of the weight $w(P_{uv})$ (respectively the length $l(P_{uv})$) of the path $P_{uv} = \{u_0, u_1 \ldots, u_n\}$ is given by the sum of the weights (respectively the* *lengths) of the arcs that compose it:*

$$w(P_{uv}) = \sum_{i=0}^{n-1} w(u_i, u_{i+1}) \quad \textit{and} \quad l(P_{uv}) = \sum_{i=0}^{n-1} l(u_i, u_{i+1})$$

Paths can be used to define connected components in the graph. The definition of connected components allows for the decomposition of the graph into subgraphs, each corresponding to a connected component.

Definition 3.8: Connected component

A connected component is a graph $G = (V, A)$ such that there exists a chain between any two vertices in $\overline{G} = (V, \overline{A})$.

A special type of a path in a graph is a cycle. In turn, the absence of cycle in a graph defines a tree, which is an important structure in graph theory. Tree structures are well-studied in graph theory and provide an efficient representation for special types of data (e.g. hierarchical data).

Definition 3.9: Cycle, tree

A cycle is a chain which starts and ends at the same vertex. A tree is a graph with no cycles.

Definition 3.10: Tree and forest

(i) *A directed tree $T = (V, A)$ rooted at vertex $r \in V$ is a directed connected component such that there exists a unique directed path from the root r to any other vertex in V.*

(ii) *An undirected tree $T = (V, A)$ is an undirected connected component without any cycle. In an undirected tree, each vertex can be considered as a root.*

(iii) *The weight $w(T)$ (respectively length $l(T)$) of the tree $T = (V, A)$ is given by the sum of the weights (respectively lengths) of the arcs in A.*

$$w(T) = \sum_{(u,v) \in A} w(u, v) \quad \textit{and} \quad l(T) = \sum_{(u,v) \in A} l(u, v)$$

(iv) *A set of trees is called a forest. The weight (respectively length) of a forest is the sum of weights (respectively lengths) of trees that compose it.*

Proposition 3.11:

Given an undirected graph $T = (V, A)$, the following properties are equivalent:

(i) T *is a tree.*

(ii) T *contains no cycle.*

(iii) T *is a connected component and* $M_T = N_T - 1$.

(iv) The removal of an arc in A *creates two components in* T*. The addition of an arc in* A *creates a cycle in* T*. In this sense, a tree is a minimum structure.*

Example: Path and tree

Consider the undirected graph $G = (V, A)$ shown in Figure 3.2(A) with weights shown on the arcs. G has two connected components G_1 and G_2, each of which is surrounded by a dashed curve for distinction.

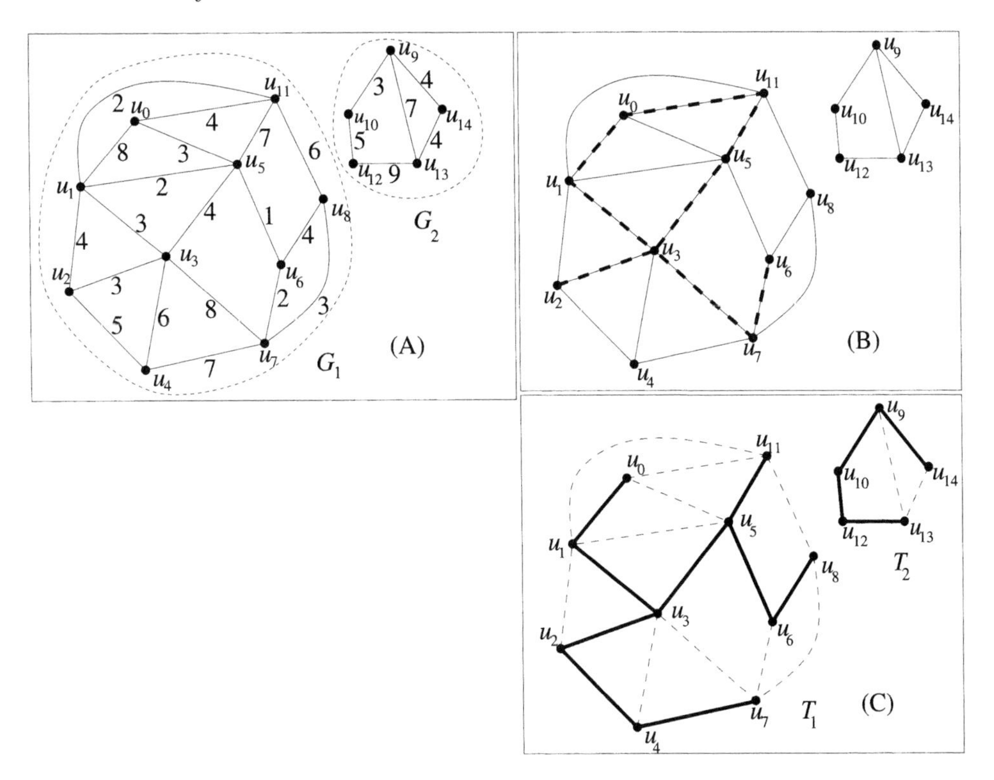

Figure 3.2 (A) Example of a graph $G = G_1 \cup G_2$ with arc lengths. (B) Non-simple path in G_1. (C) Forest $T = T_1 \cup T_2$ that spans all vertices in G

A path in G_1 from u_2 to u_6 is the sequence of vertices

$$P_{u_2 u_6} = \{u_2, u_3, u_1, u_0, u_{11}, u_5, u_3, u_7, u_6\}$$

displayed as a bold dashed line in Figure 3.2(B). Note also that the sequence $\{u_3, u_1, u_0, u_{11}, u_5, u_3\}$ forms a cycle in G_1. A sub-path of $P_{u_2u_6}$, $P_{u_2u_5} = \{u_2, u_3, u_1, u_0, u_{11}, u_5\}$ is a simple path in G_1. Given the arc weights shown in Figure 3.2(A), the weight of the path $P_{u_2u_6}$ is $w(P_{u_2u_6}) = 39$. A forest $T = T_1 \cup T_2$ that spans all vertices in G is shown as continuous lines in Figure 3.2(C). T_1 and T_2 are two trees in G_1 and G_2, respectively. The dashed lines in Figure 3.2(C) represent arcs of G that are removed from A for T_1 and T_2 to be trees. Given the arc weights in Figure 3.2(A), the weights of the two trees and the corresponding forest are, respectively $w(T_1) = 42$, $w(T_2) = 21$ and $w(T) = w(T_1) + w(T_2) = 63$.

Using the notion of path length as a distance, characteristics can be associated with a graph G.

Definition 3.12: Centre, eccentricity, radius and diameter of a connected component

Given a graph $G = (V, A)$ where A is associated with arc lengths, if G contains only one connected component then, the following parameters can be associated with G.

(i) Given a vertex $u \in V$, suppose vertex $v \in V$ is given as furthest vertex from u. In other words, the length of the shortest path from u to v is greater than or equal to the length of the shortest path from u to any vertex v' in V. Then, the length of the shortest path from u to v defines the eccentricity of u, $\mathrm{ecc}(u)$.

(ii) The centre of G is the set of vertices u of least eccentricity in G.

(iii) The radius of G is the value of the least eccentricity in G.

(iv) The diameter of G is the value of greatest eccentricity in G.

These parameters can readily be computed using the algorithms described in the next section.

3.2 Optimisation

Graphs can model formulations of optimisation problems. Common problems on graphs are the ones of locating a specific structure (e.g. a path or a tree) in the graph which optimally satisfies a given criterion. In this section, we present the solution for two such problems, namely the shortest path search (Section 3.2.1) and the minimum weighted tree search (Section 3.2.2).

3.2.1 Shortest paths

The objective of the shortest path problem is to locate a path between two given vertices whose length is minimum. This problem has numerous applications in geographical information systems and computational geometry. If no path between the two given vertices exists (e.g. they are in two different connected components in the case of an undirected graph), the algorithm would typically return a solution value of infinity, indicating that such a path does not exist.

Remark 3.13:

There may be different paths in G whose length are equal and minimal. The algorithms proposed in this section will typically find a single (not necessarily unique) optimal solution. For algorithms that enumerate all shortest paths between two vertices, the reader is referred to the relevant literature (e.g. see [49]).

Algorithmic graph theory gives a general framework for solving the shortest path problem [47]. Given a graph $G = (V, A)$ and lengths for all arcs in A, we limit the study to the case where all arc lengths are positive (i.e. $l(u, v) \geq 0$ for any $(u, v) \in A$). Given $s \in V$ and $t \in V$ as start and end vertices, respectively, the shortest path problem is that of finding a path P^*_{st} in G such that the length of this path $l(P^*_{st})$ is minimum among all possible paths P_{st} from s to t in G. Trivial properties can readily be given for the shortest path P^*_{st}.

[——— Initialisation ———]
for all vertices u in V **do**
  $d(u) \leftarrow +\infty$
$d(s) \leftarrow 0$
$\Lambda \leftarrow \{s\}$ [Initialise the list of vertices Λ with s]
[——— Main search ———]
while The list of vertices Λ is not empty **do** **(1)**
  Get and remove the vertex u from the list Λ **(2)**
  for each vertex v in the forward star of u **do**
    if $d(v) > d(u) + l(u, v)$ **then**
      $d(v) \leftarrow d(u) + l(u, v)$
      $\text{pred}(v) \leftarrow u$ [Store u as predecessor of v]
      Add v into the list of vertices Λ **(3)**
[——— Path Retrieval ———]
if $d(t) < +\infty$ **then**
  Backtrack the tree structure from t to s to get P^*_{st}
else [$d(t) = +\infty$]
  No path exists between s and t

Algorithm 3.1: The generic shortest path algorithm

Proposition 3.14:

Given a graph $G = (V, A)$, a start vertex $s \in V$ and an end vertex $t \in V$ and assuming that there exists a shortest path P^*_{st} from s to t, then,

(i) *P^*_{st} does not contain any cycle.*

(ii) *For any vertex u in P^*_{st}, there exists a shortest path P^*_{su} from s to u which is a subset of P^*_{st}.*

(iii) *For any vertex $u \in V$, $l(P^*_{st}) \leq l(P^*_{su}) + l(P^*_{ut})$. In particular, $l(P^*_{st}) = l(P^*_{su}) + l(P^*_{ut})$ for any vertex $u \in P^*_{st}$*

Typically, solutions for the shortest path problem are based on Proposition 3.14(ii) and (iii). The outline of the shortest path search method is as follows. A search which successively updates a label $d(u)$ associated with each vertex u is operated from the start vertex s. At each stage of the search, $d(u)$ is the current minimum path length found from s to u. Labels $d(u)$ are first initialised to infinity ($d(u) = +\infty, \forall u \in V$), except for the start vertex s whose label is clearly $d(s) = 0$. At some point in the search, a given vertex u is labelled with a value $d(u)$. Its adjacent vertices v are successively investigated and their respective labels $d(v)$ are updated to $\min\{d(v), d(u) + l(u, v)\}$ (Proposition 3.14(iii)). Each vertex v and its label $d(v)$ is then stored in a list Λ and the search proceeds with the next candidate vertex u in the list Λ. Paths are saved by storing the predecessor $\mathrm{pred}(u)$ of each vertex u on its path of minimum length to s. The search results in a tree structure which is the union of shortest paths from s to all other vertices in the graph. In particular, the shortest path from s to t is retrieved by backtracking from t to s on this tree. This outline is translated into the generic shortest path algorithm given in Algorithm 3.1.

This algorithm locates the shortest path P^*_{st} (whenever it exists) between vertices s and t in the graph $G = (V, A)$. The search strategy is defined by the management of the list Λ. Steps marked (2) and (3) in Algorithm 3.1 are modified according to the chosen strategy. The following part of this section details three different approaches which can be taken and presents algorithms modified according to the search strategy.

Step (1) in Algorithm 3.1 determines when the algorithm should stop the search. As given in the generic algorithm above, this stopping rule ensures an exhaustive search of all the vertices in the connected component that contains the start vertex s. Alternative stopping rules are also presented in this section.

3.2.1.1 D'Esopo-Pape's algorithm

This shortest path algorithm is based on the fact that each time a label $d(u)$ is updated (decreased) it is worth trying to decrease labels of vertices in the forward star of u (depth-first strategy), except for the first time when u is met (a breadth-first strategy is then used) [47].

```
[——— Initialisation ———]
for all vertices u in V do
   d(u) ← +∞
   η(u) ← (i)                                    [ u has never been in Λ ]
d(s) ← 0
Λ ← {s}                        [ Initialise the list of vertices Λ with s ]
η(s) ← (ii)                                                 [ s is in Λ ]
[——— Main Search ———]
while Λ ≠ ∅ do                      [ the list of vertices Λ is not empty ]
   u is the first vertex in Λ
   Λ ← Λ \ {u}                                         [ Remove u from Λ ]
   η(u) ← (iii)                [ (1)u was in Λ but has been removed ]
   for each vertex v in the forward star of u do
      if d(v) > d(u) + l(u, v) then
         d(v) ← d(u) + l(u, v)
         pred(v) ← u                       [ Store u as predecessor of v ]
         if η(u) = (i) then                      [ v has never been in Λ ]
            Λ ← Λ ∪ {v}                        [ Insert v at the end of Λ ]
            η(v) ← (ii)                                     [ v is in Λ ]
         else if η(u) = (iii) then v was in Λ and has been removed
            Λ ← {v} ∪ Λ                      [ Insert v at the start of Λ ]
            η(v) ← (ii)                                     [ v is in Λ ]
[——— Path Retrieval ———]
if d(t) < +∞ then
   Backtrack the tree structure from t to s to get P*_st
else                                                       [ d(t) = +∞ ]
   No path exists between s and t
```

Algorithm 3.2: D'Esopo-Pape's shortest path algorithm

In this sense, D'Esopo-Pape's algorithm is a label-correcting algorithm. It uses first a breadth-first search strategy. A vertex v that has never been met in the search before is stored at the end of the list Λ and its label $d(v)$ is updated to the value of the current minimal path length from s. Later in the search, this vertex is removed from the list Λ and its forward star investigated (Step (1) in Algorithm 3.2). Then, at some point in the search, this vertex is met again and becomes candidate for insertion in Λ. It is then inserted at the start of the list and is then processed using a depth-first search strategy. Each vertex v is typically associated with a status $\eta(v) \in \{(i), (ii), (iii)\}$ which is used to determine where the vertex in question is to be inserted in the list Λ [101, 113–115]. This status corresponds to the fact that a vertex v (i) has never been stored in the list Λ, (ii) is in Λ or (iii) has been in Λ but has since been removed from Λ. From the above, Step (3) in Algorithm 3.1 is replaced with the following conditions.

(i) If vertex v has never been in Λ, place vertex v at the end of the list Λ.

(ii) If vertex v is in Λ, do nothing.

(iii) If vertex v has previously been in Λ but has since been removed, place v at the start of the list Λ.

Step (2) in Algorithm 3.1 simply consists in considering the current vertex u as the first vertex in Λ (note that u is then removed from Λ).

It is known that Algorithm 3.2 is characterised by a high worst-case complexity of order $\mathcal{O}(N.2^N)$. However, in practice, this algorithm proves to be very efficient, mainly for sparse graphs, which is the context in which it is to be used. Although the storage space needed depends on the implementation, a typical value for the space requirement of this algorithm is $4N + 2M$.

3.2.1.2 Dijkstra's algorithm

In this algorithm, the idea is to investigate the forward star of the current vertex u in Λ to be scanned which has the smallest label $d(u)$ [35, 101]. Assuming that all arc lengths are positive one can prove that, at a given stage in the search, if u has the smallest label in Λ, then $d(u)$ is the length of the shortest path from s to u. Hence, the following proposition holds.

Proposition 3.15: [47]

If $l(u, v) \geq 0$, $\forall\, (u, v) \in A$, then each vertex is removed from (and hence inserted into) Λ exactly once.

The list Λ is kept sorted according to the labels $d(v)$. A status $\eta(v)$ is associated with each vertex v which denotes whether (i) v has never been in or (ii) is in Λ (v is then termed as temporarily labelled) or (iii) v has been in Λ and is removed from it (v is termed as permanently labelled). By contrast with the previous algorithm, this strategy leads to a label-setting algorithm. More precisely, Step (3) in Algorithm 3.1 is replaced with the following conditions.

(i) If vertex v has never been in Λ, it is inserted in Λ between v_1 and v_2 such that $d(v_1) \leq d(v) \leq d(v_2)$.

(ii) If vertex v is already in Λ, it is moved between v_1 and v_2 which satisfy the same criterion as above.

(iii) If vertex v has been in Λ and has since been removed from Λ, do nothing. Its label $d(v)$ is permanent and represents the length of the shortest path from s to v.

```
[——— Initialisation ———]
for all vertices u in V do
  d(u) ← +∞
  η(u) ← (i)                                    [ u has never been in Λ ]
d(s) ← 0
Λ ← {s}                        [ Initialise the list of vertices Λ with s ]
η(s) ← (ii)                                              [ s is in Λ ]
[——— Main Search ———]
while Λ ≠ ∅ do                     [ the list of vertices Λ is not empty ]
  u is the first vertex in Λ
  Λ ← Λ \ {u}                                    [ (1)Remove u from Λ ]
  η(u) ← (iii)                                           [ u was in Λ ]
  for each vertex v in the forward star of u do
    if η(v) ≠(iii) then               [ v has never been removed from Λ ]
      if d(v) > d(u) + l(u,v) then
        d(v) ← d(u) + l(u,v)
        pred(v) ← u                     [ Store u as predecessor of v ]
        if η(v) =(i) then                      [ v has never been in Λ ]
          Insert v in Λ between v1 and v2 such that d(v1) ≤ d(v) ≤ d(v2)
                                                      [ (2)Keep Λ sorted ]
        else if η(v) =(ii) then                               [ v is in Λ ]
          Move v in Λ between v1 and v2 such that d(v1) ≤ d(v) ≤ d(v2)
                                                      [ (3)Keep Λ sorted ]
[——— Path Retrieval ———]
if d(t) < +∞ then
  Backtrack the tree structure from t to s to get P*_st
else                                                    [ d(t) = +∞ ]
  No path exists between s and t
```

Algorithm 3.3: Dijkstra's shortest path algorithm

Since Λ is sorted, Step (2) in Algorithm 3.1 again consists in considering the current vertex u as the first vertex in Λ. Vertex u is then removed from Λ and its label $d(u)$ becomes a permanent.

The assumption of positive arc lengths reduces the worst-case complexity of Dijkstra's algorithm to $\mathcal{O}(N^2)$. If G is sparse, then the complexity falls to $\mathcal{O}(M \log N)$ [49]. The typical space requirement for this algorithm is the same as that of D'Esopo-Pape's algorithm with a value of $4N + 2M$.

3.2.1.3 Dial's algorithm

This algorithm is a fast implementation of Dijkstra's algorithm [33]. It uses a set of vertex lists $\{\Lambda_i\}$, called buckets, to keep the list of temporarily labelled vertices sorted while reducing the computational effort (Steps (2) and (3) in Algorithm 3.3). Each bucket Λ_i corresponds to a different length $l(P_{su}) = i$ of a path starting from s to a subset of vertices $\{u\}$. The number of such buckets needed by the algorithm to store all the vertices is derived from the following proposition (for the sake of simplicity, we assume here that labels can only take integer values).

Proposition 3.16: [47]

Let $l_{\max}$ be the maximal arc length in G (i.e. $l_{\max} = \max\{l(u,v)\ ;\ (u,v) \in A\}$). Assuming that all arc lengths are positive, if $d(u)$ is minimal for all vertices u in the vertex list, then this list contains only vertices v such that $d(u) \leq d(v) \leq d(u) + l_{\max}$.

If at each stage, the newly updated vertex v is inserted at the end of the bucket numbered $(d(v) \bmod (l_{\max}+1))$, only $(l_{\max}+1)$ buckets $\{\Lambda_i\ ;\ i = 0, \ldots, l_{\max}\}$ are necessary for the complete process. Note that the set of buckets is first initialised empty and vertex s is added to the bucket corresponding to a zero-length path.

Here, Step (3) in Algorithm 3.1 is replaced with the following rule. Vertex v is placed in the bucket corresponding to its new label $d(v)$ (i.e. the bucket Λ_i, where $i = d(v) \bmod (l_{\max}+1)$). Step (2) in Algorithm 3.1 consists in considering the current vertex u as the first vertex in the first non-empty bucket Λ_k (i.e. the bucket corresponding to the the smallest current path length). Buckets are inspected in a circular fashion starting from the index k of the bucket Λ_k where the previous vertex u was removed. If, during this inspection, all $(l_{\max} + 1)$ buckets are visited and empty, one can conclude that all the vertices in the connected component of the graph containing s have been visited (Step (3) in Algorithm 3.4). In this case, the search stops and the shortest path from s to any other vertex can be retrieved by backtracking on the tree structure.

The use of buckets allows the algorithm to readily keep the list of vertices sorted without the need for searching where to insert a new vertex. In this case, the complexity of the algorithm is $\mathcal{O}(l_{\max}N + M)$. Clearly, this algorithm is efficient only if $l_{\max}$ is small compared with M or N. The typical storage space needed is $5N + 2M + l_{\max}$.

3.2.1.4 Shortest path spanning tree and stopping rule

All algorithms described above result in a tree that spans all vertices in the connected component of G that contains the start vertex s. This spanning tree is formed by the union of all shortest paths from s to all vertices of this connected component. Such a tree will be referred to as the shortest path spanning tree rooted at s.

```
[——— Initialisation ———]
for all vertices u in V do
   d(u) ← +∞
for i = 0 to l_max do                                   [ All buckets are empty ]
   Λ_i ← ∅
d(s) ← 0
Λ_0 ← {s}    [ s is stored in the bucket corresponding to a zero-length path ]
k ← 0                                               [ The search starts from s ]
[——— Main search ———]
while k ≥ 0 do                         [ (1)While there is a non-empty bucket ]
   u is the first vertex in Λ_k
   Λ_k ← Λ_k \ {u}                              [ (2)Remove u from bucket Λ_k ]
   for each vertex v in the forward star of u do
      if d(v) > d(u) + l(u, v) then
         d(v) ← d(u) + l(u, v)
         pred(v) ← u                              [ Store u as predecessor of v ]
         i = d(v) mod (l_max + 1)
         Λ_i ← Λ_i ∪ {v}
            [ Insert v at the end of the bucket numbered d(v) mod (l_max + 1) ]
   i ← 0
   while Λ_k = ∅ and i ≤ l_max + 1 do  [ Find the first non-empty bucket... ]
      k ← (k + 1) mod (l_max + 1)                       [ ...in a circular fashion ]
      i ← i + 1
   if i > l_max + 1 then                               [ (3)All buckets are empty ]
      k ← −1               [ The main while loop (Step (1) above) terminates ]
[——— Path Retrieval ———]
if d(t) < +∞ then
   Backtrack the tree structure from t to s to get P*_st
else                                                            [ d(t) = +∞ ]
   No path exists between s and t
```

Algorithm 3.4: Dial's shortest path algorithm

Remark 3.17:

D'Esopo-Pape's algorithm (Algorithm 3.2) proves to be very efficient in practice when the complete shortest path spanning tree of a sparse connected component is needed.

During shortest path algorithms, the shortest path spanning tree is stored by associating a predecessor $\text{pred}(u)$ with each vertex $u \in V$. In the special case of the start vertex s, s is considered as its own predecessor since it terminates any shortest path in the tree structure. In other words, the shortest path spanning tree is rooted at s. As a result of the algorithms given above, the length of the shortest path from s to any vertex $u \in V$ is given by the final label $d(u)$. The backtracking procedure which finds the optimal shortest path from s to any other vertex u in the component (e.g. the end vertex t, if s and t are connected) given the shortest path spanning tree is presented in Algorithm 3.5.

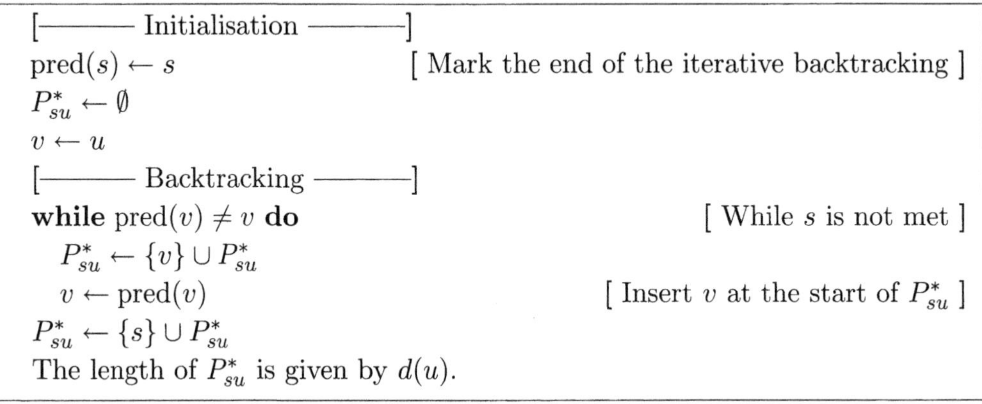

[——— Initialisation ———]
$\text{pred}(s) \leftarrow s$ [Mark the end of the iterative backtracking]
$P^*_{su} \leftarrow \emptyset$
$v \leftarrow u$
[——— Backtracking ———]
while $\text{pred}(v) \neq v$ **do** [While s is not met]
  $P^*_{su} \leftarrow \{v\} \cup P^*_{su}$
  $v \leftarrow \text{pred}(v)$ [Insert v at the start of P^*_{su}]
$P^*_{su} \leftarrow \{s\} \cup P^*_{su}$
The length of P^*_{su} is given by $d(u)$.

Algorithm 3.5: Backtracking procedure in the shortest path spanning tree

In the case where only the shortest path (and its length) from s to t is needed, it is important that the shortest path search stops as soon as possible, in order to avoid unnecessary computations. In the case of label-setting algorithms (e.g. Dijkstra's and Dial's algorithms presented above as Algorithms 3.3 and 3.4, respectively), the search strategy used allows for ending the shortest path search as soon as the given end vertex t is reached. With the aid of Proposition 3.15, it is clear that the label of a vertex u becomes permanent once u is removed from the vertex list Λ. Therefore, Algorithms 3.3 and 3.4 can be modified in order to end the search as soon as the label $d(t)$ becomes permanent (i.e. Steps (1) and (2) in Algorithms 3.3 and 3.4, respectively).

Moreover, in both cases, all vertices whose labels are smaller than $d(t)$ have permanent labels. They are therefore eligible for ensuring a correct backtracking as given in Algorithm 3.5. This property will prove fundamental in image processing applications where there is a need for a topological ordering of vertices. Both Algorithms 3.3 and 3.4 implicitly create this ordering by expanding their search boundary as a "disc" around the start vertex s. At any stage in these algorithms, all vertices v which have a label smaller than that of the vertex u whose forward star is under investigation are permanently labelled. Moreover, in the case of Dial's algorithm, buckets readily form sets of vertices of equal shortest path lengths from s. This will also prove useful for some digital image processing applications.

Finally, heuristics exist for modifying the order in which vertices u are investigated. This approach for guiding the search for the shortest path was first introduced in [55, 108]. More recently [156], a heuristic called "oval-search strategy", based on the prior knowledge of the positions of the start and end vertices, was presented. Instead of extending the search boundary in a symmetric fashion around the start vertex, this heuristic directs the search towards the end vertex. The efficiency of this heuristic on geographical networks is demonstrated in [156].

3.2.2 Minimum weighted spanning tree

In this section, the solution to another classic graph-theoretic optimisation problem is presented. Given a graph $G = (V, A)$ which forms a single connected component and arc weights for all arcs in A, the problem of finding a minimum weighted spanning tree in G is that of finding a tree $T^* = (V, A_{T^*})$ that spans all vertices in G and whose weight is minimum among all possible such trees T. More formally, given G, the minimum weighted spanning tree T^* in G satisfies the following conditions.

(i) T^* is a tree and spans all vertices in G (i.e. $T^* = (V, A_{T^*})$).

(ii) $w(T^*) = \min\{w(T)$; T satisfies condition (i) above $\}$.

The minimum weighted spanning tree problem finds many applications in problems where a set of locations is to be connected at minimum cost (e.g. telecommunication networks).

Remark 3.18:

There is a clear difference between the concept of a shortest path spanning tree of a graph and that of a minimum weighted spanning tree. The shortest path spanning tree is such that the length of any path from the start vertex s to any other given end vertex t is minimal. The minimum weighted spanning tree minimises the sum of the weight of the arcs it includes.

Two classic algorithms are presented here for the solution of the minimum weighted spanning tree problem. Given a graph $G = (V, A)$, both algorithms construct a minimum weighted spanning tree $T = (V, A_T)$ in G as follows. Starting with $A_T = \emptyset$, arcs are successively added to A_T with regard to a given criterion. This criterion is typically based on the following two propositions and the fact that a tree $T = (V, A_T)$ that spans all vertices in $G = (V, A)$ should include exactly $M_T = N_G - 1$ arcs (see Proposition 3.11).

Proposition 3.19: [49]

Given a graph $G = (V, A)$ and arc weights, consider the tree $T = (V, A_T)$ which spans all vertices of G. Given an arc (u, v) in $A \setminus A_T$ (i.e. (u, v) is not in the *tree T), the addition of (u,v) in A_T creates a cycle, P_{uu} say (see Proposition 3.11). Then, if for any arc (s,t) in this cycle $w(u,v) \geq w(s,t)$, T is a minimum weighted spanning tree in G.*

In other words, Proposition 3.19 states that T is a minimum weighted spanning tree if and only if, in any cycle created by the addition of an extra arc (u,v) in A_T, the arc of highest weight in this cycle is (u,v).

Proposition 3.20: [49]

Given a graph $G = (V, A)$ and arc weights, consider the tree $T = (V, A_T)$ which spans all vertices of G. If, for any arc (u,v) in A_T (i.e. any arc (u,v) which is part of the tree T), the set $A' \subset A \setminus A_T$ of arcs (s,t) which are not included in A_T and have a vertex in common with (u,v) is such that $w(s,t) \geq w(u,v)$, $\forall\, (s,t) \in A'$ then, T is a minimum weighted spanning tree in G

Proposition 3.20 characterises a minimum weighted spanning tree as a tree T that spans all vertices in G and such that any arc which is part of T has a weight smaller than or equal to any of its adjacent arcs which are not included in T.

3.2.2.1 Kruskal's algorithm

This algorithm [80] follows the construction method suggested by Proposition 3.19. Arcs in A are first sorted in increasing order of their weights. At each stage of the tree construction, the set $A_T^{\complement}$ contains arcs of A that are not included in A_T. Then, arcs are successively added to A_T (i.e. included in the tree T) in the order given by $A_T^{\complement}$, provided that they do not form a cycle in T (see Proposition 3.19). The construction stops when $(N_G - 1)$ arcs have been added to A_T (i.e. when the tree T contains $(N_G - 1)$ arcs).

```
[——— Initialisation ———]
Order arcs in A in ascending order of weight (1)
A_T ← ∅
M_T ← 0
A_T^C ← A
[——— Construction ———]
while M_T < N_G − 1 do
   (u,v) is the first arc in A_T^C        [ (u,v) is the arc of minimum weight ]
   A_T^C ← A_T^C \ {(u,v)}                 [ Remove the arc (u,v) from A_T^C ]
   if (u,v) does not create a cycle in T then (2)
      A_T ← A_T ∪ {(u,v)}                   [ Add the arc (u,v) in A_T ]
      M_T ← M_T + 1
```

Algorithm 3.6: Kruskal's minimum weighted spanning tree algorithm

Clearly, the sub-graph $T = (V, A_T)$ is a tree since the condition that T does not contain any cycle is maintained during its construction. Moreover, at each stage, the tree constructed contains arcs of minimum possible weights. Therefore, at each stage, T satisfies the condition given in Proposition 3.19 (relative to the sub-graph that contains the vertices currently spanned by T). Therefore, the tree T thus constructed is a minimum weighted spanning tree in G.

The complexity of Algorithm 3.6 mostly depends on the implementation of Steps (1) and (2). Sorting the arcs (Step (1)) can optimally be done in $\mathcal{O}(M_G \log M_G)$ operations. At each stage, one should check whether the arc (u, v) does not close a cycle in T (Step (2)). This operation can clearly be achieved in $\mathcal{O}(M_T)$ operations. Now, $(N_G - 1)$ such steps are to be performed. Therefore, (Step (2) can be achieved in $\mathcal{O}(N_G \log N_G)$. Hence, the overall complexity of Algorithm 3.6 is given by Step (1) as $\mathcal{O}(M_G \log M_G)$.

3.2.2.2 Prim's algorithm

This algorithm uses a different construction strategy [121]. It is based on Proposition 3.20 and the fact that the tree should span all the vertices in V.

[——— Initialisation ———]
for all vertices u in V **do**
  $\text{pred}(u) \leftarrow \emptyset$ [All vertices are disconnected]
  $\pi(u) = +\infty$
$V_T^{\mathsf{c}} \leftarrow V$
$A_T \leftarrow \emptyset$
$\pi(s) \leftarrow 0$
[——— Construction ———]
while $V_T^{\mathsf{c}} \neq \emptyset$ **do**
  u is the vertex in V_T^{c} with minimal potential $\pi(u)$ (1)
  $V_T^{\mathsf{c}} \leftarrow V_T^{\mathsf{c}} \setminus \{u\}$ [Remove u from V_T^{c}]
  if $\text{pred}(u) \neq \emptyset$ **then**
    $A_T \leftarrow A_T \cup \{(\text{pred}(u), u)\}$
  for all vertex $v \in V_T^{\mathsf{c}}$ in the forward star of u **do** (2)
    if $w(u, v) < \pi(v)$ **then** [Update vertex v]
      $\pi(v) \leftarrow w(u, v)$
      $\text{pred}(v) \leftarrow \{u\}$

Algorithm 3.7: Prim's minimum weighted spanning tree algorithm

A root s is arbitrarily chosen (see Definition 3.10(ii)). A value $\pi(v)$ (potential of v) and an adjacent vertex $\text{pred}(v)$ are associated with each vertex $v \in V$. $\pi(v)$ represents the weight of the arc $(\text{pred}(v), v)$ which has made it possible to connect s to v (i.e. $(\text{pred}(v), v)$ is the last arc on the path from s to v in the tree). At each stage, V_T^c stores vertices associated with a temporary potential. Starting with $V_T^c = V$, vertices u in V_T^c are successively considered in the order of increasing potential. The vertex u with the minimum temporary potential $\pi(u)$ is selected and its potential becomes permanent. The arc $(\text{pred}(u), u)$ is then added to A_T and u is removed from V_T^c. Before proceeding with the next vertex u in V_T^c, vertices v in the forward star of u which have a temporary potential $\pi(v)$ are updated as follows. If v is such that $w(u, v) < \pi(v)$, then $\pi(v)$ is updated to $w(u, v)$ and $\text{pred}(v)$ to $\{u\}$. The algorithm stops when all potentials have become permanent (i.e. when $V_T^c = \emptyset$). Clearly, at each stage of the construction, the tree T satisfies the condition given by Proposition 3.20. Hence, Algorithm 3.7 results in a minimum weighted spanning tree of G.

The complexity of this algorithm arises from the fact that there are N_G vertices (u's) to be removed from V_T^c. Each of these removal operations requires $\mathcal{O}(N_G)$ operations (e.g. Steps (1) and (2)). Therefore, the overall complexity of Algorithm 3.7 is $\mathcal{O}(N_G^2)$. This complexity can be reduced to $\mathcal{O}(M_G \log N_G)$ in the case of sparse graphs [49].

3.3 Analogies with digital image processing

Algorithmic graph theory clearly operates on discrete data and it is, therefore, well-suited for the type of data encountered in digital image processing. Moreover, graph-theoretic concepts such as vertex, arc, path and shortest path length can readily be mapped onto digital image processing concepts such as pixel, neighbourhood, digital arc and discrete distance. Other analogies are also presented which will be used when dealing with digital image processing applications in latter chapters.

Section 3.3.1 first illustrates how a digital image can readily be mapped onto a graph. From then on, a number of analogies map directly by comparing results and definitions presented in Chapters 1 and 2 with those presented in this chapter. Section 3.3.2 extends these analogies to the concept of discrete distance presented in Section 1.4.

3.3.1 Image-to-graph mapping

A graph $G = (V, A)$ is based on the definition of a set of discrete data (vertices in V) and their inter-relationship (arcs in A). As described in Chapter 1, a digital image is a set of discrete points on which a digital topology can be defined. Moreover, digital topology introduces the concept of neighbourhood for a pixel which, in turn, defines digital arcs and curves (see Section 1.3.1).

It is therefore clear that a graph $G = (V, A)$ can be defined using the set of pixels F in the image as set of vertices V. Such a graph is referred to as the grid graph of the image.

Definition 3.21: Grid graph [89, 133, 152]

Given a set of pixels F in the image and a connectivity relationship on which a digital topology is based, the grid graph $G = (V, A)$ of the image is defined as follows.

(i) To every pixel p in F there corresponds a vertex u in V.

(ii) An arc (u, v) exists in A whenever the pixels p and q corresponding to vertices u and v, respectively, are neighbours in the digital topology.

(iii) The length associated with the arc (u, v) is the length of the move made between the corresponding two pixels p and q, respectively.

(iv) The abstract grid graph corresponding to the infinite lattice is called the complete grid graph.

Immediate properties of the grid graph are given in Proposition 3.22.

Proposition 3.22:

By definition of digital topology (see Chapter 1),

(i) The grid graph $G = (V, A)$ of an image is generally sparse. The number of pixels that are neighbours to a given pixel is limited by the size of this neighbourhood. Typically, $M_G \leq k.N_G$, where $k = 4$ (4-neighbourhood), 8 (8-neighbourhood) or 16 (16-neighbourhood).

(ii) Lengths of the arcs in the grid graph are generally given by small integers (e.g. from Section 1.4, typical values are ($a = 3$, $b = 4$) or ($a = 5$, $b = 7$, $c = 11$)).

(iii) The grid graph on a set of N pixels can be constructed in linear time (i.e. in $\mathcal{O}(N)$ operations).

Remark 3.23:

In Chapters 1 and 2, pixels were identified with discrete points (pixel centres). From now on, a further analogy identifies pixels and vertices in the grid graph. Therefore, pixels will be equivalently referred to as discrete points (e.g. p, q) or vertices (e.g. u, v). Similarly, depending on the context, the set of pixels will be equivalently denoted F or V, by analogy with the set of vertices in the grid graph. Finally, arcs in the grid graph will be equivalently referred to as moves on the underlying lattice.

The notion of connectivity between pixels is mapped onto that of adjacency between vertices in the grid graph. Therefore, both definitions of connected components in digital topology and graph theory are clearly equivalent.

Remark 3.24:

Note that spanning tree algorithms presented in Section 3.2 can be used to identify connected components in the grid graph. Moreover, since the grid graph is sparse, the D'Esopo-Pape algorithm (Algorithm 3.2) will be particularly efficient for this task (see Remark 3.17). This process is detailed in Section 6.1.2 and compared with other classical approaches.

Moreover, using the image-to-graph mapping, the concept of neighbourhood of a pixel is directly mapped onto that of the forward star of a vertex. In the case of a complete grid graph, the forward star of a vertex u readily contains the neighbourhood of the corresponding pixel p (e.g. $N_8(p)$). In the case where the grid graph spans only vertices corresponding to a subset F of pixels in the image (e.g. the foreground pixels in the image), the forward star of a vertex u in such a grid graph will characterise the pixels that are neighbours to u and which are included in F (e.g. $N_8(p) \cap F$).

Example: Grid graph of the foreground of a binary digital image

Consider the binary digital image shown in Figure 3.3(A). The set F of foreground pixels (i.e. black pixels) is displayed as black circles ($\bullet$). Empty circles ($\circ$) represent background pixels (i.e. white pixels) in F^{c}. Figure 3.3(B) shows the grid graph of the foreground F when considering the 8-neighbourhood relationship (i.e. 8-grid graph). Clearly, this graph is sparse.

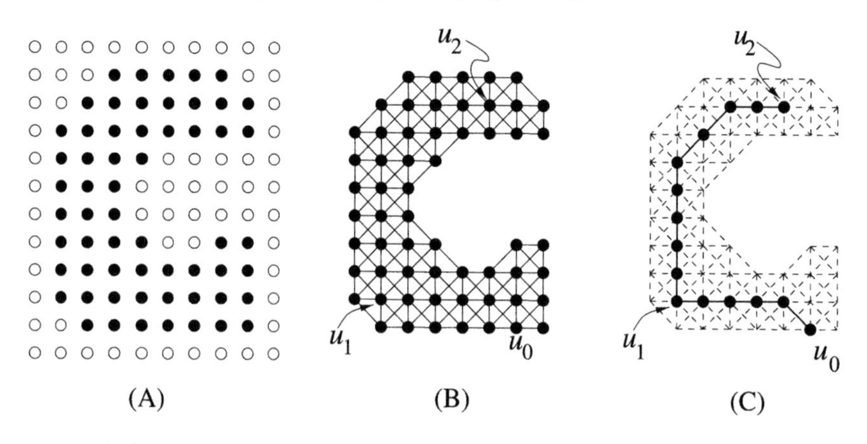

Figure 3.3 (A) Binary digital image. (B) Corresponding 8-grid graph. (C) A path in the grid graph

In this example, F is considered as a closed set and therefore border pixels are foreground pixels (i.e. $\Gamma \subset F$). By definition of an interior pixel p (i.e. $p \notin \Gamma$), all dual neighbours of p are included in F (i.e. $N_4(p) \subset F$ in this case, see Definition 1.11). Therefore, such a pixel p is characterised in the grid graph by a vertex whose forward star contains $|N_4(p)| = 4$ vertices v such that

$l(u, v) = a$. For example, this is the case for vertices u_1 and u_2 in Figure 3.3(B). By contrast, u_0 is a border vertex.

Figure 3.3(C) shows an example of a path $P_{u_0u_2}$ between vertices u_0 and u_2 in the grid graph. The analogy between such a path and a digital arc is discussed next.

The mapping between a digital arc and a path in the grid graph needs further precision. A digital arc was defined as a set of neighbouring pixels such that each pixel in the digital arc has exactly two neighbours, except for the start and end vertices (see Definition 1.4). It is therefore clear that a corresponding path in the grid graph is a simple path. However, an additional condition for a simple path in the grid graph to correspond to a digital arc in the image is required. This condition simply states that, for each vertex u in such a path, exactly two vertices in the forward star of u in the graph are included in the path, except for the start and end vertices, each of which has only one adjacent vertex in the path in question. This condition will always be satisfied on any shortest path in a grid graph as discussed in the next section.

For example, in Figure 3.3(C), $P_{u_0u_2}$ does not correspond to an 8-digital arc since the predecessor of vertex u_1 has three 8-neighbours on this path. However, it is easy to verify that each sub-path $P_{u_0u_1}$ and $P_{u_1u_2}$ defines an 8-digital arc.

3.3.2 Shortest paths and discrete distances

The grid graph defined in the previous section allows for the formulation of optimisation problems in digital images. Referring to Section 3.2, a basic optimisation problem is that of locating a shortest path. By analogy with the definition of a discrete distance given in Definition 1.18, it is clear that the discrete distance value between two pixels is the length of the shortest path between the two corresponding vertices in the grid graph [54, 89, 97, 99, 102, 154] (see Remarks 1.19 and 3.13). Moreover, the properties of a shortest path given in Proposition 3.14 justify the fact that such a length defines a distance (see also Definition 1.15).

Such an analogy allows for the use of efficient shortest path algorithms presented in Section 3.2.1. Moreover, the grid graph is well-suited for the use of these algorithms since it is a sparse graph with typically small positive integer arc lengths (see Proposition 3.22). Finally, using the graph-theoretic approach one can take advantage of by-products arising from these algorithms. For example, when calculating the shortest path spanning tree using Dial's algorithm (Algorithm 3.4), discrete discs are readily defined by vertices in buckets. Dial's shortest path algorithm can therefore readily be used to expand discrete discs in the grid graph (see Chapter 5).

Definition 3.25: Shortest path base graph

Given the grid graph $G = (V, A)$ *with arc lengths and two vertices* $u \in V$ *and* $v \in V$*, the shortest path base graph associated with the vertices* u *and* v *is the subgraph* $\mathrm{SPBG}(u, v)$ *of* G *formed by all possible shortest paths from* u *to* v.

Remark 3.26:

Following the convention established in Chapter 1, the notation for the shortest path base graph will include the dependency of the neighbourhood relationship considered with an index k *(i.e.* SPBG_k*) corresponding to that neighbourhood space (e.g.* $k = 4, 8, 16$*).*

Properties are associated with the shortest path base graph. Typical properties of a shortest path base graph in the 8-neighbourhood space are given in the example below.

Example: Shortest path base graph in the 8 neighbourhood space (SPBG_8)

Consider the complete 8-grid graph $G = (V, A)$ presented in Figure 3.4(A). Given the two vertices $u, v \in V$, the shortest path base graph $\mathrm{SPBG}_8(u, v)$ is shown as bold lines in Figure 3.4(B).

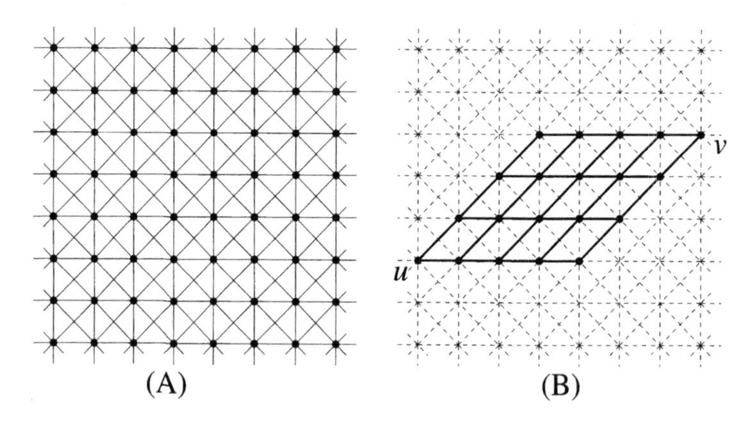

Figure 3.4 (A) 8-Neighbourhood complete grid graph. (B) Shortest path base graph $\mathrm{SPBG}_8(u, v)$

Montanari [99] has proved that there exists a shortest path in the complete 8-grid graph between any two vertices $u, v \in V$ which consists of only two straight segments, one horizontal (or vertical) and one diagonal. It is therefore clear that any shortest path in the complete 8-grid graph will be composed of at most two basic directions. Hence, the shortest path base graph $\mathrm{SPBG}_8(u, v)$ is included in a parallelogram shape, as shown in Figure 3.4(B).

The analogy between discrete distances and lengths of shortest paths in the complete grid graph is summarised in Definition 3.27 and Proposition 3.28.

Definition 3.27: Number of moves in a path

In the 8-grid graph, arcs correspond to either a- or b-moves. In this respect, given an 8-path P_{uv} *between two vertices* u *and* v, $k_{\mathrm{a}}(u,v)$ *(respectively* $k_{\mathrm{b}}(u,v)$*) denotes the number of arcs corresponding to a-moves (respectively b-moves) in* P_{uv}*. The length of* P_{uv} *is therefore given by* $l(P_{uv}) = a.k_{\mathrm{a}}(u,v) + b.k_{\mathrm{b}}(u,v)$.

Similarly, in the 16-grid graph, the length of a 16-path P_{uv} *is given by* $l(P_{uv}) = a.k_{\mathrm{a}}(u,v) + b.k_{\mathrm{b}}(u,v) + c.k_{\mathrm{c}}(u,v)$*, where* $k_{\mathrm{c}}(u,v)$ *is the number of arcs corresponding to c-moves in* P_{uv}.

Moreover, the following proposition forms the basis for the calculation of discrete and continuous distances using shortest path lengths.

Proposition 3.28:

In a complete grid graph, the following properties hold.

(i) Given two vertices u *and* v *in the complete 8-grid graph* $G = (V, A)$*, if the shortest 8-path* P_{uv} *in* G *from* u *to* v *is of length* $l(P_{uv}) = a.k_{\mathrm{a}}(u,v) + b.k_{\mathrm{b}}(u,v)$*, then* $d_{a,b}(p,q) = a.k_{\mathrm{a}}(u,v) + b.k_{\mathrm{b}}(u,v)$*, where* p *and* q *are the points that correspond to vertices* u *and* v*, respectively. Moreover, the Euclidean distance* $d_{\mathrm{E}}(p,q)$ *is calculated as*

$$d_{\mathrm{E}}(p,q) = \sqrt{(k_{\mathrm{a}}(u,v) + k_{\mathrm{b}}(u,v))^2 + k_{\mathrm{b}}(u,v)^2}$$

(ii) Given two vertices u *and* v *in the complete 16-grid graph* $G = (V, A)$*, the shortest 16-path* P_{uv} *in* G *from* u *to* v *is composed of two types of arcs only. Moreover, if* $l(P_{uv}) = a.k_{\mathrm{a}}(u,v) + b.k_{\mathrm{b}}(u,v) + c.k_{\mathrm{c}}(u,v)$ *and* p *and* q *are the points corresponding to vertices* u *and* v*, respectively, then,* $d_{a,b,c}(p,q) = a.k_{\mathrm{a}}(u,v) + b.k_{\mathrm{b}}(u,v) + c.k_{\mathrm{c}}(u,v)$ *and*

$$d_{\mathrm{E}}(p,q) = \sqrt{(k_{\mathrm{b}}(u,v) + k_{\mathrm{c}}(u,v))^2 + (k_{\mathrm{a}}(u,v) + k_{\mathrm{b}}(u,v) + 2k_{\mathrm{c}}(u,v))^2}$$

The 8-grid graph considered now is that shown in Figure 3.3. By contrast with a complete grid graph, it is the grid graph of a bounded connected component. Figure 3.5 highlights the difference between the shortest path spanning tree and the minimum weighted spanning tree. Figure 3.5(A) shows the shortest path spanning tree rooted at u_0 obtained with arc lengths $a = 3$ and $b = 4$.

In such an 8-grid graph, the previous description of a shortest path is not always valid since the shortest path between two vertices may be constrained by the border of the component. Figure 3.5(B) shows such a shortest path of length $l(P_{u_0u_2}) = 31$ between u_0 and u_2. Clearly, this shortest path in the grid graph corresponds to an 8-digital arc. Moreover, it is the only possible shortest path

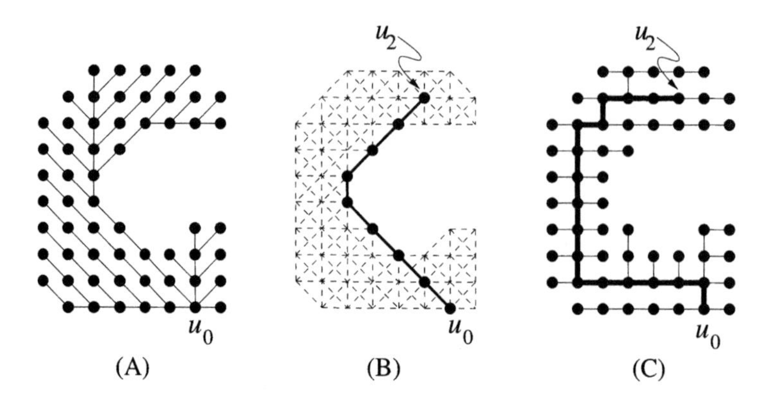

Figure 3.5 (A) Shortest path spanning tree in the grid graph shown in Figure 3.3(A). (B) An example of shortest path. (C) Minimum weighted spanning tree in the same grid graph

between u_0 and u_2 in the grid graph. Figure 3.5(C) shows a minimum weighted spanning tree obtained when using arc lengths as arc weights. Clearly, with such weights, the minimum spanning tree will be composed of a-moves only. This tree should contain $(|F|-1)$ arcs since it connects $N = |F|$ vertices (see Proposition 3.11). Now, $a < b$. Therefore, the minimum possible weight for a tree T that spans all vertices in F is $w(T) = (|F|-1)a$. This tree is to be contrasted with the shortest path spanning tree shown in Figure 3.5(A). In particular, note that the unique path $P_{u_0u_2}$ in the minimum weighted spanning tree is not a shortest path. In Figure 3.5(C), $P_{u_0u_2}$ is highlighted as a bold line and $l(P_{u_0u_2}) = 51$.

Further geometrical analogies can be defined between graph theory and digital image processing. For example, the analogy between a shortest path and a digital straight segment is described in detail in [89, 152]. This analogy corresponds to the intuitive definition which can be given to a digital straight segment.

Proposition 3.29: [89,152]

(i) A shortest path in a grid graph is a digital arc in the digital topology induced by the set of pixels spanned by the grid graph in question.

(ii) A digital straight segment in the 8- or 16-neighbourhood is a shortest path in the corresponding complete grid graph.

Proof:

(i): The proof is given in the case of an 8-grid graph (other cases are similar). Given a shortest path $P_{u_0u_n}$ in the 8-grid graph, if $P_{u_0u_n}$ is not a digital arc, then there exists a vertex u_i which has three 8-neighbours u_j, u_k and u_l in $P_{u_0u_n}$. Clearly, for every combination of order in which vertices u_i, u_j, u_k and u_l appear in $P_{u_0u_n}$, a sub-path of $P_{u_0u_n}$ can be defined between u_0 and u_n

by creating a "shortcut" such that vertex v_i has only two 8-neighbours in this sub-path. This contradiction highlights the fact that the original path $P_{u_0 u_n}$ is a shortest path. Therefore, Proposition 3.29(i) holds.

(ii): From the above description of the shortest path base graph, it is clear that any shortest path in the complete 8-grid graph satisfies Freeman's two first conditions for discrete straightness given in Proposition 2.4. Moreover, since the grid graph is assumed to be complete, by definition of the shortest path base graph, all possible combinations of these two moves are present in the shortest path base graph. Therefore, it is clear that there exists at least one shortest path that satisfies the third condition in Proposition 2.4. In that case, this shortest path is a digital straight segment on the infinite lattice. In Section 2.2.1, it was shown that all possible digital straight segments between two points are equivalent via shift operations. Since the shortest path base graph contains any possible combinations of the two moves present in this digital straight segment, the polygon defined by the upper and lower bounds of this digital straight segment is totally included in the shortest path base graph. Therefore, all possible digital straight segments between the two points in question are included in the shortest path base graph. Therefore, Proposition 3.29(ii) holds. By analogy, the case of the 16-neighbourhood is trivial (see Section 2.2.3). □

Remark 3.30:

Note that Proposition 3.29(ii) implies that all possible digital straight segments which exist between two given points have the same minimal length.

In summary, the grid graph derived from any lattice (i.e. triangular, hexagonal or square lattice), readily contains all information about neighbourhood relationships and distances between pixels. Moreover, using characterisations of geometrical and topological properties such as straightness or convexity, well-known optimisation algorithms (e.g. see Section 3.2) can be used to perform optimally digital image processing operations (e.g. see [89, 152]).

The graph-theoretic approach will therefore often form the basis for further developments of digital image processing operators. Throughout this book, this approach will be consistently developed for both low-level image processing (pixel level) and intermediate-level processing image (component level).

Chapter 4

ACQUISITION AND STORAGE

The mapping from a continuous to a discrete image forms the first step in any digital image processing application. Discrete data resulting from this digitisation process is then stored in a form which is suitable for further processing. In this book, we are mostly concerned with binary digital image processing and, hence, binary image acquisition. Data resulting from the acquisition process is typically composed of black and white pixels, represented by integer points on a lattice associated with 0-1 values.

Section 4.1 reviews in detail the acquisition step already briefly introduced in the previous chapters. The study of partitions which model acquisition devices and their dual lattices presented in Chapter 1 is first briefly summarised. Then, different digitisation methods are presented and their usage is justified. In particular, the square-box quantisation and the grid-intersect quantisation are formally re-introduced and compared.

The problem of image data storage is considered in the second part of this chapter. Depending on the approach chosen for further processing, the storage will vary in order to facilitate access to data throughout the process. In this respect, different methods for image data storage are presented in Section 4.2. The form in which image data is stored and the approach chosen for the processing naturally define the type of data-structures which are to be used in this context. Such data-structures are also presented in this section. Section 4.2 terminates with an introduction to binary digital image data compression. Simple techniques are presented here.

4.1 Digitisation

In Chapter 1, three types of regular partitions of the real plane were presented and studied, leading by duality to the triangular, square and hexagonal lattices on which the pixel centres lie, respectively. In Section 1.4.4, it was shown that the hexagonal lattice had limited practical applications. The triangular lattice arises from the hexagonal partition, which is an accurate model of physical acquisition devices. However, because of the simplicity of the square lattice and the fact that the triangular lattice can easily be mapped onto it, square lattices have wide applications and have generated most of theoretical developments. In this section, we therefore concentrate on the acquisition process using square partitions which, in turn, define square lattices.

Digitisation is the process of mapping a set S of real points onto a set P of discrete points p_i (i.e. black pixels). S is referred to as the continuous set and P is referred to as the digitisation set of S. In the case of the digitisation of a continuous curve C, C is given an orientation which will define an ordering on the points in P. The following terminology, extended from that presented in [37,67] will be used throughout.

Definition 4.1: Preimage and domain of a digitisation set P

(i) Given a set P of discrete points, a continuous set S whose digitisation is P is called a preimage of P.

(ii) The continuous area defined by the union of all possible preimages of P is called the domain of P.

Remark 4.2:

This terminology was first introduced as follows:

(i) In [67], the preimage of a set P is a continuous set S whose square-box quantisation is P (see Section 4.1.2). By contrast, this definition applies here to any specified digitisation scheme.

(ii) In [37], the domain of a digital straight segment defined using the object-boundary quantisation is characterised as a minimal polygon. We first extend this terminology to any digitisation scheme and simply consider the domain as given by Definition 4.1 (i.e. no attempt is made to characterise the minimal set that includes all preimages of a digitisation set P).

Since the square lattice is used in this chapter, a pixel p_i is assumed to have integer coordinates (x_i, y_i). This is always possible via the definition of the sampling step (see Definition 4.3). Moreover, the use of other lattices can readily be emulated by the square lattice via straightforward calculations [169].

The approach for creating a method for the acquisition process is described in this section. Four digitisation schemes will be presented. The first scheme is only used as an example to highlight the characteristics which should be fulfilled by a satisfactory digitisation scheme. Then, three different digitisation techniques based on the partition and its dual lattice, respectively, are presented.

More precisely, the characterisation of a satisfactory digitisation technique is as follows. A trivial technique is first presented in Section 4.1.1. This method and the inconsistencies it leads to are detailed and illustrated by examples. Based on this, the characterisation of an acceptable generic digitisation technique is given. The definition of a digitisation method which is equivalent to the square-box quantisation briefly introduced in Section 2.2.2 is given as one such acceptable digitisation method. This technique and its properties are presented in Section 4.1.2. A different approach for the digitisation of thin continuous objects such as curves leads to the definition of the grid-intersect quantisation, also briefly introduced in Section 2.2.1. This grid-intersect quantisation is studied in more detail in Section 4.1.3. These two digitisation techniques are commonly used in digital image processing. Another important digitisation scheme, called object-boundary quantisation is based on a third approach and is presented in Section 4.1.4. Its similarities with the grid-intersect quantisation are shown for some special instances of a continuous set S. Finally, Section 4.1.5 summarises the most important features of these three digitisation schemes.

4.1.1 A simple digitisation scheme

A trivial digitisation scheme is presented here. The square lattice is first superimposed on the continuous object S and a pixel p represented by a point on the square lattice belongs to the digitisation of S if and only if $p \in S$. A general example for this scheme is given in Figure 4.1, where the square lattice is represented as dashed lines and S is illustrated by the shaded area. Pixels in the digitisation set P of S are shown as black circles ($\bullet$) (all other pixels are white and are omitted for clarity).

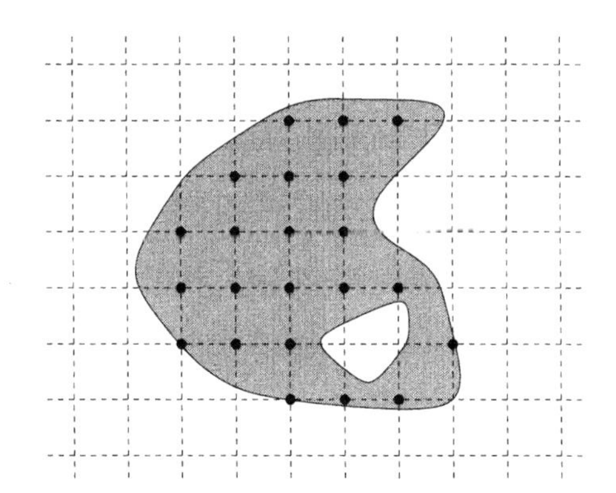

Figure 4.1 A simple digitisation scheme

Any digitisation process relies on the definition of a sampling step which creates a relationship between the discrete and continuous scales.

Definition 4.3: Sampling step

The sampling step $h > 0$ is the length of the side of a partition polygon. In the special case of the square lattice, the sampling step is by definition the real value $h > 0$ defining the distance value between two 4-neighbouring pixels.

The sampling step $h > 0$ acts as a scaling factor when digitising continuous sets, hence allowing for different resolutions. Clearly, the smaller the value that h takes, the higher is the resolution. An example is shown in Figure 4.2.

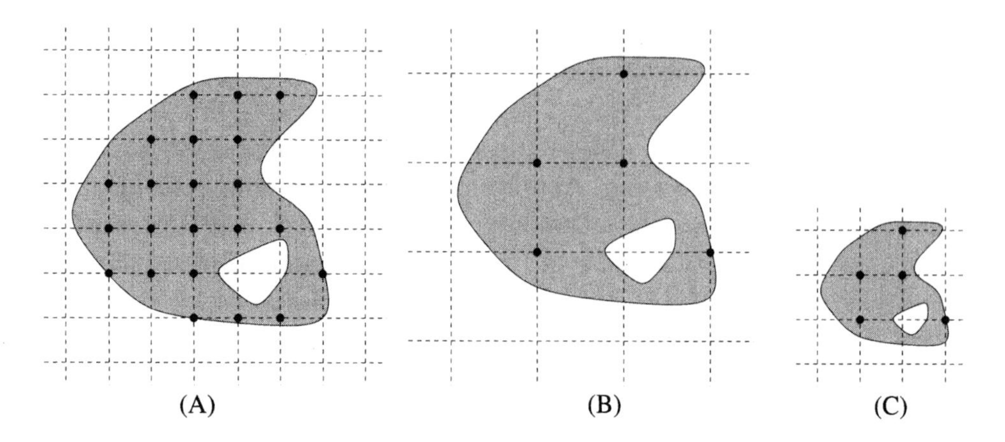

Figure 4.2 Digitisation of a continuous set S using different sampling step values. (A) Sampling step h. (B) Sampling step $h' = 2 \times h$. (C) Digitisation of the same continuous set using a scale factor $\frac{h}{h'} = \frac{1}{2}$ and a sampling step h

In Figure 4.2(A), the continuous set S is digitised with a given sampling step h. Figure 4.2(B) displays the digitisation of the same set S using a sampling step $h' = 2 \times h$. Clearly, this process is equivalent to resizing the continuous set with a scale factor $\frac{h}{h'}$ and using h as the sampling step (see Figure 4.2(C)).

The analysis of this digitisation scheme includes different aspects. In summary, the following problems will be highlighted and illustrated by examples.

(i) A non-empty set S may be mapped onto an empty digitisation set.

(ii) This digitisation scheme is non-invariant under translation.

(iii) Given a digitisation set P, the characterisation of its possible preimages S is imprecise.

Since a pixel p appears in the digitisation set of a continuous set S if and only S contains p, any continuous set S in which all integer points p_i are removed leads to an empty digitisation set as shown in Figure 4.3(A). Similarly, it is easy to define a non-empty set S that does not contain any integer point. Figure 4.3(A) also emphasises the fact the probability for a thin object such as a straight line or a curve to include an integer point is very low. In this sense, thin objects will generally lead to empty discrete sets P using this digitisation method. The non-invariance of the digitisation set under translation of the continuous set S is clearly shown in Figure 4.3(B). The "C"-shaped set S may be mapped onto an empty, disconnected, or connected digitisation set, respectively, depending on its position on the lattice. The variation of P under translation of S is referred to as "aliasing" in image processing terminology.

Finally, Figure 4.3(C) suggests that continuous objects of radically different shapes can lead to the same digitisation. This is explained by the fact that

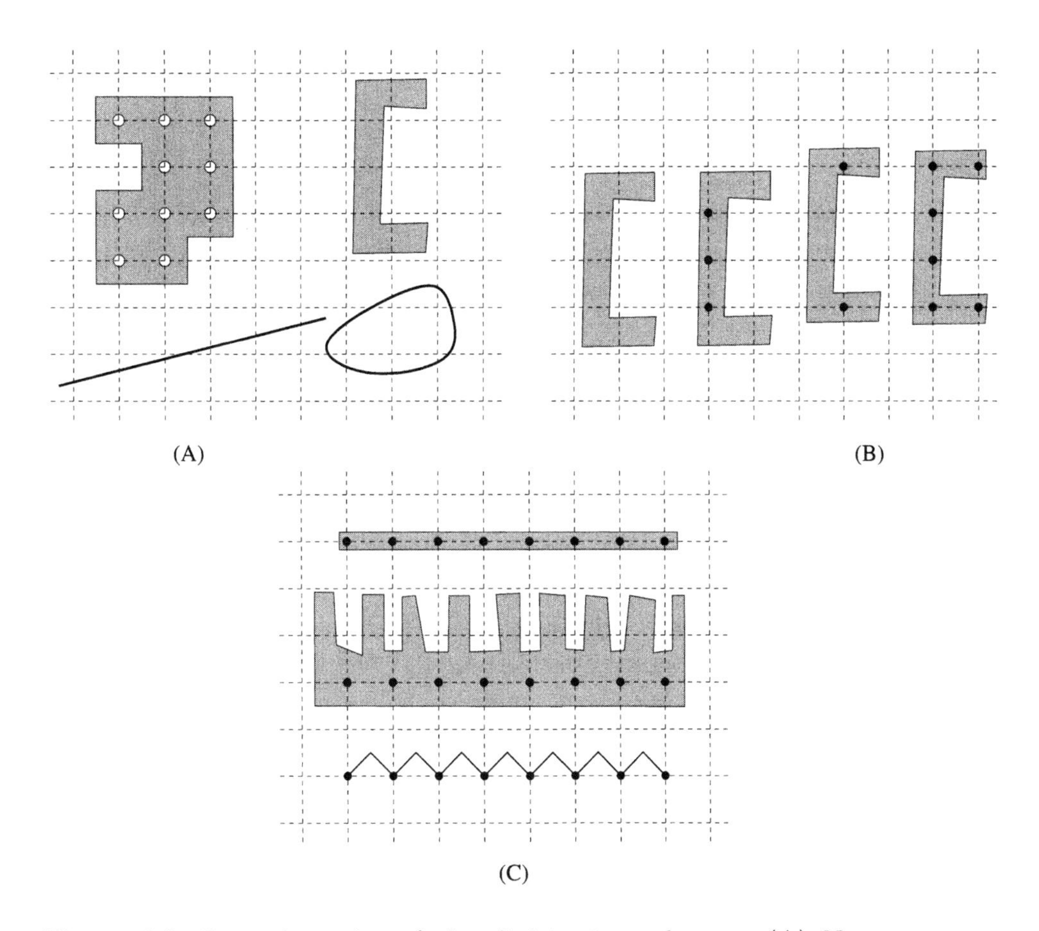

Figure 4.3 Inconsistencies of the digitisation scheme. (A) Non-empty sets mapped onto empty digitisation sets. (B) Aliasing. (C) Examples of continuous sets that can be associated with a digitisation set

the domain of a digitisation set P is not bounded. This is again in contradiction with the compatibility of discrete geometrical properties such as convexity or straightness with their continuous counterparts (see Chapter 2).

Characteristics of an acceptable digitisation scheme are therefore given by Proposition 4.4.

Proposition 4.4:

An acceptable method for the digitisation scheme should include the following features

(i) The digitisation of a non-empty continuous set is non-empty.

(ii) The digitisation scheme should be reasonably invariant under translation (i.e. the aliasing should be reduced as much as possible).

(iii) Given a digitisation set P, preimages of P should be found similar regarding

a certain criterion. More formally, the domain of P should be bounded and as small as possible.

Remark 4.5:

Clearly, one can overcome all problems summarised in Proposition 4.4 by decreasing the sampling step as much as necessary. However, this solution is not always possible since, in practical applications, the resolution of the digitisation process is limited by the characteristics of the acquisition device. Therefore, it is of interest to overcome or, at least, diminish these problems without requiring an unrealistic resolution.

In Section 4.1.2, the first approach used to overcome these problems is based on the complete coverage of the continuous set S by areas associated with pixels (referred to as digitisation boxes). Section 4.1.3 presents another scheme which is designed for the digitisation of thin objects such as curves or straight segments. Sets of pixels obtained from this digitisation technique are shown to be eligible for the study of geometrical properties (see Chapter 2). The third method presented in Section 4.1.4 is used for digitising boundaries of continuous sets.

4.1.2 Square-box quantisation

Using the digitisation scheme presented in Section 4.1.1, only integer points of the continuous set S are represented in the digitisation set P of S. The approach presented in this section aims to associate a unique pixel p_i in the digitisation set P with any real point γ in the continuous set S in question. Clearly, using such a digitisation method, any non-empty continuous set S will be mapped onto a non-empty digitisation set P since every real point is associated with a discrete point.

This digitisation method is described as follows. A digitisation box $B_i =]x_i - \frac{1}{2}, x_i + \frac{1}{2}[\times]y_i - \frac{1}{2}, y_i + \frac{1}{2}[$ is associated with any pixel $p_i = (x_i, y_i)$. Digitisation boxes are typically equivalent to partition polygons whose centres are the pixels, as defined in Chapter 1. Then, a pixel p_i appears in the digitisation set P of S if and only if $B_i \cap S \neq \emptyset$ (i.e. its associated digitisation box B_i is intersected by S).

Figure 4.4 illustrates the resulting digitisation set of the set S presented in Figure 4.1. Black pixels in P are again shown as black discs ($\bullet$). White pixels are not shown for clarity. Dotted lines represent the square partition, dual to the square lattice. Graphically, a pixel p_i appears in the digitisation set if S intersects its corresponding partition polygon.

Using the digitisation method thus defined, any real point $\gamma \in B_i$ is mapped onto a unique pixel p_i. The case of real points γ which lie on horizontal or vertical lattices lines ($y = y_i + \frac{1}{2}$) and ($x = x_i + \frac{1}{2}$), respectively is now

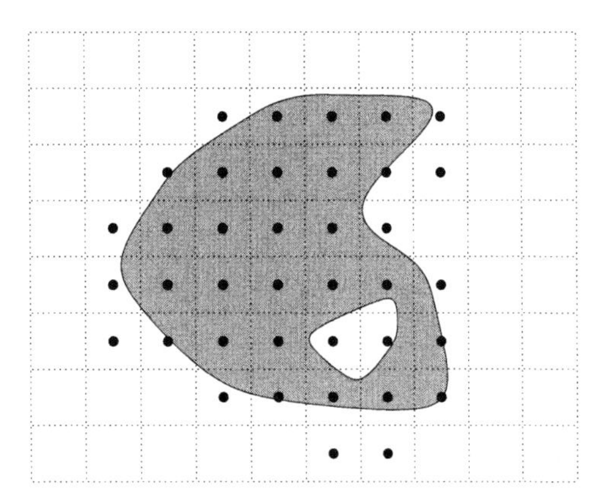

Figure 4.4 Digitisation based on digitisation boxes

to be considered since such points are not associated with any point in P. The definition of digitisation boxes is first modified as follows.

Definition 4.6: Semi-open tile quantisation [22, 44]

Let $B_i = [x_i - \frac{1}{2}, x_i + \frac{1}{2}[\times[y_i - \frac{1}{2}, y_i + \frac{1}{2}[$*, the digitisation set* P *of a continuous set of points* S *is the set of pixels* $\{p_i$ *such that* $B_i \cap S \neq \emptyset\}$*. More formally, each point* $\gamma = (x_\gamma, y_\gamma) \in S$ *is mapped onto* $p_i = (x_i, y_i) \in P$ *such that* $x_i = \text{round}(x_\gamma)$ *and* $y_i = \text{round}(y_\gamma)$.

Clearly, using Definition 4.6, any real point γ is mapped onto a unique pixel p_i on the square lattice. However, as illustrated in Figure 4.5, the digitisation of a continuous diagonal straight segment $[\alpha, \beta]$ (i.e. the slope σ of $[\alpha, \beta]$ is such that $|\sigma| = 1$) is not necessarily a 4- or an 8-digital arc, which is in contradiction with the study of discrete straightness presented in Chapter 2. The digitisation box used for the semi-open tile quantisation is displayed as a shaded area. Alternative continuous and dashed borders of this area highlight the fact that the digitisation is a semi-open square. The dashed broken line illustrates the order in which pixels in P are obtained when digitising $[\alpha, \beta]$.

Therefore, the digitisation scheme is modified as detailed in Definition 4.7. This leads to the square-box quantisation used in Chapter 2 when studying geometrical properties of 4-connected sets.

Definition 4.7: Square-box quantisation, SBQ [67, 135] ("cellular image", according to [67])

Let $B_i =]x_i - \frac{1}{2}, x_i + \frac{1}{2}[\times]y_i - \frac{1}{2}, y_i + \frac{1}{2}[$*. The closure of* B_i *is* $\overline{B}_i = [x_i - \frac{1}{2}, x_i + \frac{1}{2}] \times [y_i - \frac{1}{2}, y_i + \frac{1}{2}]$*. A point* p_i *appears in the digitisation set* P *of a continuous set* S *if and only if* $B_i \cap S \neq \emptyset$*. Moreover, in the case of a continuous curve* C*, for any real point* $\gamma \in C$ *such that* γ *lies exactly on a horizontal or vertical*

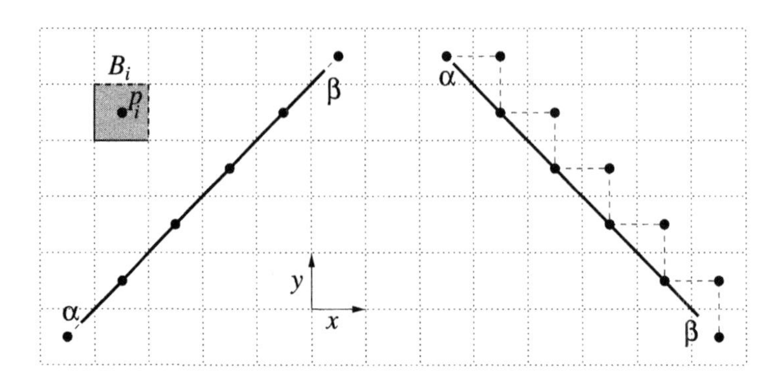

Figure 4.5 Semi-open tile quantisation of a continuous straight segment

lattice line, there exist two digitisation boxes B_i and B_j such that $\gamma \in \overline{B}_i \cap \overline{B}_j$. The pixel of p_i and p_j which is locally at the left of C is selected to be in the digitisation of C.

Remark 4.8:

The choice of the pixel which is "locally at the left of C" is arbitrary, as is the choice of the orientation of C.

It can easily be shown that the square-box quantisation of a connected set S is 4-connected (see the proof of Proposition 4.9, next). This scheme therefore eases the study of geometrical characteristics of the resulting digitisation set (see Chapter 2). Moreover, the following proposition highlights the advantage of the square-box quantisation on the semi-open tile quantisation.

Proposition 4.9: [67, 135]

The square-box quantisation of a continuous straight segment is a 4-digital arc.

Proof:
The 4-connectedness of P is shown in the general case where C is a continuous curve. Let p_i be a pixel in P, the digitisation set of C. Then, there exists at least one real point $\gamma \in C$ such that $\gamma \in B_i$. Following C along its orientation, for "exiting" B_i, C has to intersect the border of B_i at some point γ'. If $\gamma' \in C \cap B_i$ is not a corner of B_i, by continuity of C, the curve C enters the digitisation box B_{i+1} corresponding to a pixel p_{i+1} which is a 4-neighbour of p_i. Since $p_{i+1} \in P$, P is 4-connected in this case. If $\gamma' \in C \cap B_i$ is a corner of B_i, the choice of the pixel locally at the left of C ensures that p_{i+1} is also a 4-neighbour of p_i in this case. Figure 4.6(A) enumerates all possible cases (other cases are equivalent via symmetry or rotation) of such a point γ'. Therefore, the digitisation set P of a continuous curve S is 4-connected.

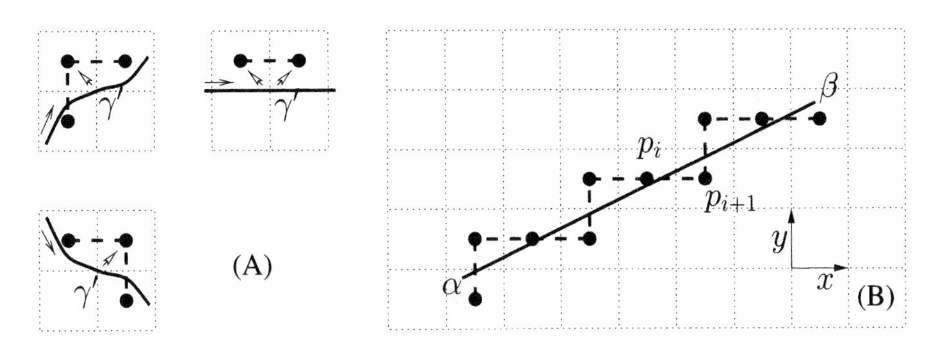

Figure 4.6 (A) Special cases for the mapping of real points onto pixels. (B) Square-box quantisation of the continuous segment $[\alpha, \beta]$ (the slope σ of $[\alpha, \beta]$ is such that $-1 < \sigma < 1$)

Now, let $C = [\alpha, \beta]$ be a continuous segment oriented from α to β and such that its slope is positive (i.e. $\sigma \geq 0$, other cases are equivalent by symmetry). Let $p_i = (x_i, y_i)$ and $p_{i+1} = (x_{i+1}, y_{i+1})$ be two successive pixels in P (see Figure 4.6(B)). Since p_i and p_{i+1} are 4-neighbours, $d_4(p_i, p_{i+1}) = |x_{i+1} - x_i| + |y_{i+1} - y_i| = 1$. Moreover, since $[\alpha, \beta]$ is a straight segment and $\sigma \geq 0$, increments $(x_{i+1} - x_i)$ and $(y_{i+1} - y_i)$ are clearly monotonic and positive. Therefore, at each step, $(x_{i+1} - x_i) + (y_{i+1} - y_i) = 1$. Hence, P is a 4-digital arc. □

The definition of the square-box quantisation also ensures that a non-empty set S is mapped onto a non-empty digitisation set P. Similarly, it is easily seen that the reduction of aliasing is preserved by this modification.

Remark 4.10:

The definition of the square-box quantisation does not make it suitable for implementation in the case where the set S is given via an analytical definition (e.g. corresponding to a function of $\mathbb{R}^2$). The choice of the pixel "locally at the left of C" involves the orientation of C, which, in turn, implies a global understanding of the input set S.

However, if each of the square partition polygons whose centre lies on the dual square lattice is associated with a physical captor, digitisation techniques based on digitisation boxes are the most accurate and straightforward techniques to implement. In this context, the probability for a border point of S to match exactly with the border of a partition polygon is negligible. In other words, this case is mostly considered for the completeness of the theoretical model which is used for the study of cellular properties (see Chapter 2 for more details on cellular straightness and convexity).

The class of digitisation schemes based on digitisation boxes is now studied and their properties are summarised and illustrated by examples.

The use of digitisation boxes ensures that any real point γ is mapped onto a unique discrete point. This, in turn allows for any non-empty continuous set to be mapped onto a non-empty digitisation set. Clearly, complete invariance under translation cannot be achieved. However, aliasing is greatly reduced using such a digitisation scheme. This is illustrated in Figure 4.7 where the continuous set used in Figure 4.3 is used again.

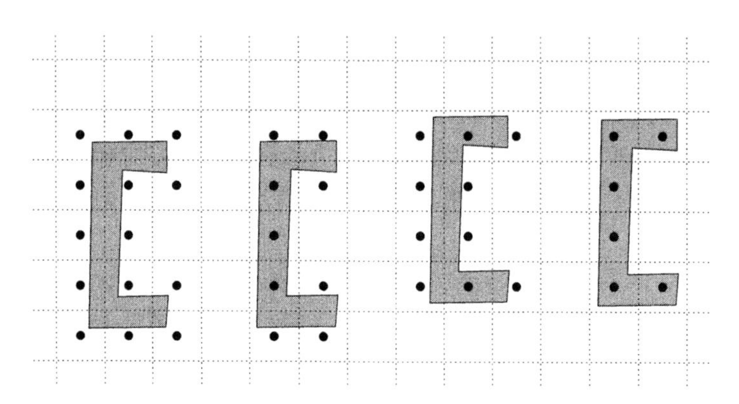

Figure 4.7 Reduction of aliasing with the use of digitisation boxes

Moreover, the square-box quantisation ensures properties on sets of discrete points resulting from the digitisation of continuous sets. As noted earlier, such a scheme has been used to defined cellular properties.

Finally, the use of digitisation boxes allows for restricting the set of preimages of P. Given a set of discrete points P, any SBQ-preimage of P is such that $S \subseteq \bigcup\{B_i \text{ such that } p_i \in P\}$. As illustrated in Figure 4.8, the definition of the square-box quantisation improves the quality of the estimation made by the first definition (compare with Figure 4.3(C)). Firstly, the SBQ-domain of P is included in the union of digitisation boxes associated with pixels in P. This set is illustrated as a shaded area in Figure 4.8(A) (the dashed border emphasises the fact that, in the general case, this set is open). Moreover, S is an SBQ-preimage of P if and only if S intersects all digitisation boxes associated with pixels in P. Figures 4.8(B) and (C) display examples of such sets S. The dashed polygon recalls the border of the shaded area shown in Figure 4.8(A).

In particular, any continuous straight segment that intersects all digitisation boxes defined by P is mapped onto P. Therefore, the set of SBQ-preimages of P are similar in shape. In other words, P is an accurate representative of its SBQ-preimages. However, the following shortcomings remain.

(i) The number of connected components in the background of a picture may not be preserved by the square-box quantisation. In other words, holes in S may not appear in P (e.g. see Figure 4.4).

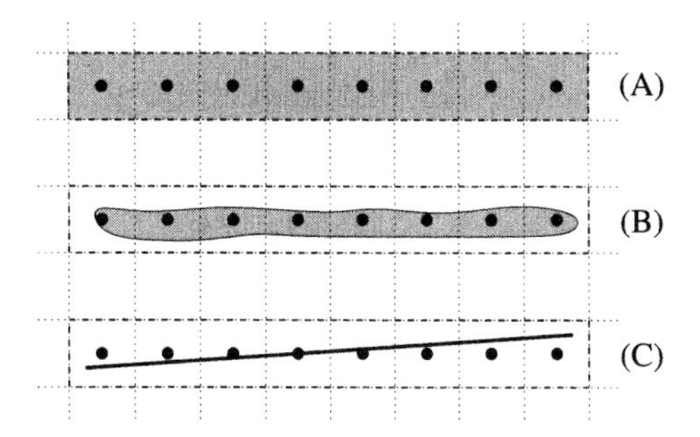

Figure 4.8 Examples of SBQ-preimages of a digitisation set P (compare with Figure 4.3). (A) SBQ-domain of P. (B) and (C) Plausible preimages of P

(ii) Similarly, since $S \subset \cup B_i$, the number of connected components ν_S in a continuous set S can only be related to the number of connected components ν_P in its corresponding digitisation set P by the inequality $\nu_P \leq \nu_S$ (see Figure 4.9(A))

(iii) More generally, no relation can be made between the digitisation set P of a continuous set S and the digitisation set P' of S^c, the complementary set of S (see Figure 4.9(B)).

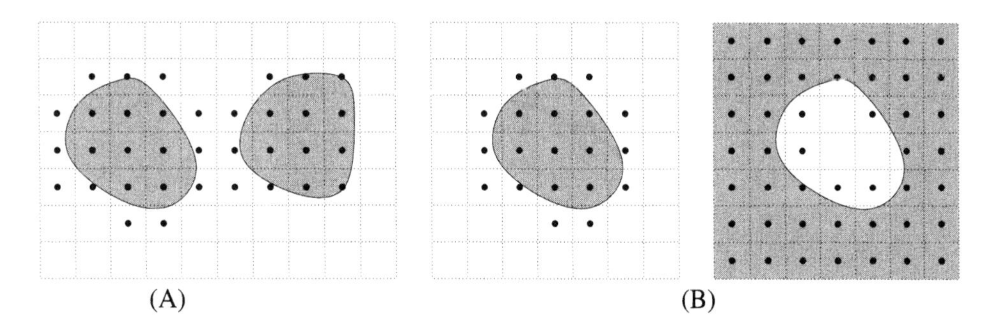

Figure 4.9 Inconsistencies of the digitisation scheme. (A) Non-conservation of the number of connected components. (B) Digitisation of a set and its complement

No digitisation scheme exists which overcomes all these shortcomings. The solution adopted in general is to adapt the resolution for the digitisation to the size of details in S (i.e. adapt the value of the sampling step, see Definition 4.3 and Figure 4.2).

4.1.3 Grid-intersect quantisation

This digitisation scheme is designed for the mapping of continuous thin objects such as segments or curves onto sets of discrete points. It is based on the intersection of the continuous object C in question and the lattice lines. The square lattice is used again here. The definition of the grid-intersect quantisation sketched in Section 2.2.1 is recalled more formally.

Definition 4.11: Grid-intersect quantisation, GIQ [43, 135, 137]

Given a continuous thin object C, each intersection of C with a lattice line defines a real point $\gamma = (x_\gamma, y_\gamma)$ such that either $x_\gamma \in \mathbb{Z}$ or $y_\gamma \in \mathbb{Z}$ depending on whether C intersects a vertical or horizontal lattice line, respectively. Such a point $\gamma \in C$ is mapped onto a lattice point $p_i = (x_i, y_i)$ such that $x_\gamma \in]x_i - \frac{1}{2}, x_i + \frac{1}{2}[$ and $y_\gamma \in]y_i - \frac{1}{2}, y_i + \frac{1}{2}[$. In the special case of a tie (i.e. $x_\gamma = x_i + \frac{1}{2}$ or $y_\gamma = y_i + \frac{1}{2}$), the discrete point p_i which is locally to the left of C is selected to be in the digitisation set P.

Remark 4.12:

As in the square-box quantisation, the choice of the orientation of C is arbitrary.

The grid-intersect quantisation of a generic curve C is illustrated in Figure 4.10. Points in P (i.e. black pixels) are shown as black discs (•). White pixels are not shown for clarity. In this example, C is oriented clockwise, as illustrated by the arrow. Any intersection point between C and the small continuous segments leading to highlighted pixels is mapped onto the corresponding pixel.

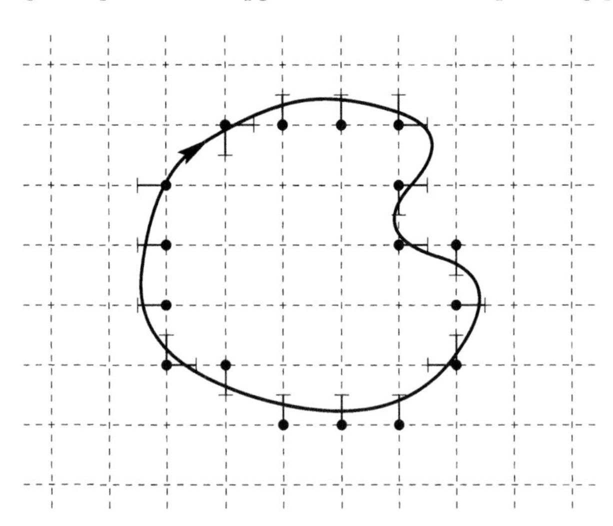

Figure 4.10 Grid-intersect quantisation of a continuous curve C

Clearly, P is 8-connected if C is a continuous connected set (see proof of Proposition 4.13, next). In the general case of a curve C, P is not guaranteed to be a 4- or 8-digital arc or closed curve. However, one can prove that the grid-intersect quantisation of a continuous straight segment is an 8-digital arc.

Proposition 4.13: [137]

The grid-intersect quantisation of a continuous straight segment $[\alpha, \beta]$ is an 8-digital arc. Moreover, it can be defined using intersections between $[\alpha, \beta]$ and either horizontal or vertical lines only, depending on the slope of $[\alpha, \beta]$.

Proof:
It is first shown that the grid-intersect quantisation P of a generic continuous curve C is 8-connected (Figure 4.10 shows an example where P is not 4-connected). Let p_i be a pixel in P. p_i is defined by the intersection of C with the border of a lattice square adjacent to p_i. Therefore, there exists a real point $\alpha \in C$ that belongs to the interior of this lattice square. For "exiting" this lattice square, C should intersect again its border. In any case, this intersection is mapped onto a pixel p_{i+1}, 8-neighbour of p_i (a special case is that of $p_i = p_{i+1}$). Therefore, P is 8-connected.

Now, let $C = [\alpha, \beta]$ be a continuous straight segment such that its slope σ is positive (other cases are equivalent by symmetry). Let $p_i = (x_i, y_i)$ and $p_{i+1} = (x_{i+1}, y_{i+1})$ be two successive pixels in P. Clearly increments in coordinates are such that $(x_{i+1} - x_i) \geq 0$ and $(y_{i+1} - y_i) \geq 0$. Since p_i and p_{i+1} are 8-neighbours, $\max(|x_{i+1} - x_i|, |y_{i+1} - y_i|) = 1$. Therefore, for any p_i in P, $\max((x_{i+1} - x_i), (y_{i+1} - y_i)) = 1$. Hence, P is an 8-digital arc. Figure 4.11 illustrates the grid-intersect quantisation of a continuous straight segment $[\alpha, \beta]$.

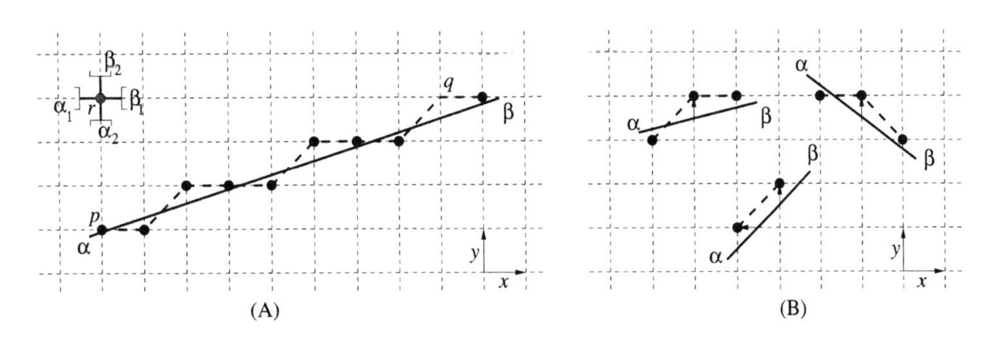

Figure 4.11 Digitisation scheme for the continuous segment $[\alpha, \beta]$ (the slope σ of $[\alpha, \beta]$ is such that $-1 < \sigma < 1$). (A) General case. (B) Special cases where $[\alpha, \beta]$ intersects a lattice line exactly half-way between two pixels

Let us now assume that $|\sigma| < 1$ and let $\gamma = (x_\gamma, y_\gamma)$ be the intersection point between $[\alpha, \beta]$ and the vertical lattice line $(y = y_i)$. Let $p_i = (x_i, y_i)$ be the pixel that represents γ in P. In other words, $y_\gamma = y_i$ and $|x_i - x_\gamma| < \frac{1}{2}$ (the case of equality is trivial by enumeration and is not detailed here). Let $\gamma' = (x_{\gamma'}, y_{\gamma'})$ be the intersection point between $[\alpha, \beta]$ and the horizontal lattice line $(x = x_i)$. Therefore, $x_{\gamma'} = x_i$. By definition, $|\sigma| = \frac{|y_{\gamma'} - y_\gamma|}{|x_{\gamma'} - x_\gamma|}$. Substituting with the above values, one obtains $|y_{\gamma'} - y_i| = |\sigma| \times |x_i - x_\gamma|$. Since $|\sigma| < 1$ and $|x_i - x_\gamma| < \frac{1}{2}$, $|y_{\gamma'} - y_i| < \frac{1}{2}$. In other words, γ' is also mapped onto p_i by the grid-intersect quantisation. More generally, if $|\sigma| < 1$, intersections with horizontal lattice lines are mapped onto pixels already defined by intersections between $[\alpha, \beta]$ and vertical lattice lines. Therefore, P can be defined using intersections between $[\alpha, \beta]$ and vertical lattice lines only (e.g. see Figure 4.11). □

Remark 4.14:

As pointed out in Section 4.1.2 in the case of the semi-open tile quantisation, a digitisation scheme similar to the grid-intersect quantisation in which an intersection point $\gamma = (x_\gamma, y_\gamma)$ is mapped onto a pixel $p_i = (\text{round}(x_\gamma), \text{round}(y_\gamma))$ does not necessarily guarantee that the digitisation of a continuous straight segment $[\alpha, \beta]$ is a 4- or 8-digital arc (see Figure 4.12(A)).

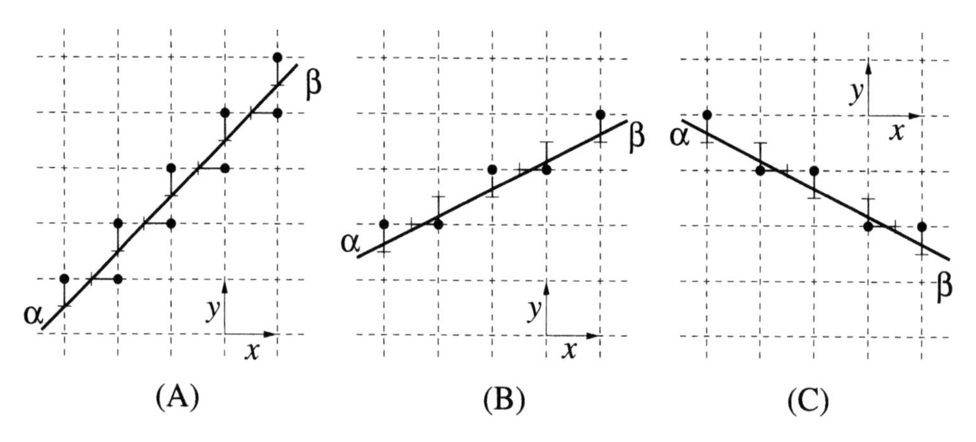

Figure 4.12 Grid-intersect quantisation of a continuous straight segment. (A) Using the round(.) function, the digitisation may not be a 4- or 8-digital arc. Here, $\sigma = 1$ and $\mu = k + \frac{1}{2}$ for some $k \in \mathbb{N}$. (B) and (C) The use of the rounding function may depend on the slope and the orientation of $[\alpha, \beta]$

It can, however, easily be shown that, depending on the slope of $[\alpha, \beta]$, such a rounding scheme can be used and results in the grid-intersect quantisation of $[\alpha, \beta]$. For example, the round(.) *function can be used for the continuous segment $[\alpha, \beta]$ having a slope $0 \le \sigma < 1$ (e.g. see Figure 4.12(B), where $\sigma = \frac{1}{2}$). Proposition 4.13 shows that the grid-intersect quantisation of $[\alpha, \beta]$ is completely defined by the pixels $p_i = (x_i, y_i)$ such that $y_i = \text{round}(y_\gamma)$, where $\gamma = (x_i, y_\gamma)$ is the intersection between $[\alpha, \beta]$ and the vertical lattice line $(x = x_i)$. Other cases can be found similarly by symmetry or rotation of angle $\frac{\pi}{2}$ (e.g. see Figure 4.12(C)).*

The study of the grid-intersect quantisation is based on the same criteria as before (see Proposition 4.4). Firstly, following Remark 4.10, it should be noted that the grid-intersect quantisation is mostly used as a theoretical model for the acquisition process. As shown earlier, it is designed to produce 8-connected digitisation sets on which discrete geometrical properties which generally correspond to their continuous counterparts can be studied (see Chapter 2).

Clearly, the grid-intersect quantisation of a continuous object C is non-empty if and only if C intersects at least one lattice line. In other words, continuous objects strictly contained in a lattice square defined by four 4-connected pixels are considered as negligible. This shortcoming is readily resolved by choosing an appropriate sampling step for the digitisation of C (see Definition 4.3).

The aliasing created by the grid-intersect quantisation is illustrated in Figure 4.13. In the case where the value of sampling step h is appropriate for the digitisation of C, this aliasing is similar to that created by the square-box quantisation.

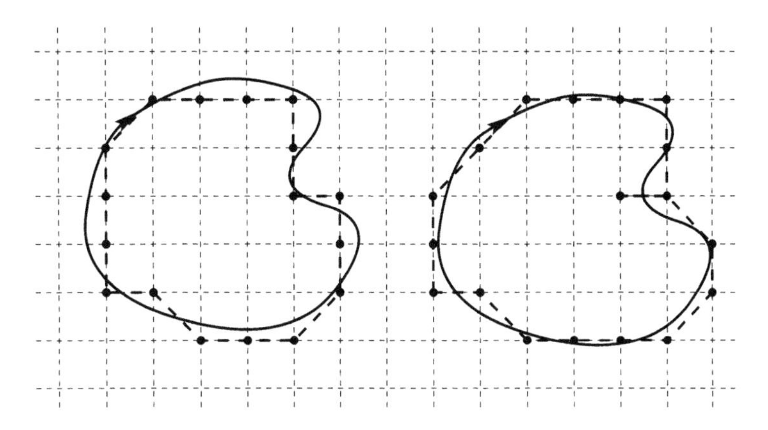

Figure 4.13 Aliasing created by the grid-intersect quantisation

The characterisation of GIQ-preimages of P is carried out as follows. Each pixel $p_i \in P$ is associated with the continuous subset which is mapped onto it. This subset is illustrated in Figure 4.14(A). Let p_j be a pixel which is 4-connected with $p_i \in P$ and $p_{j+\frac{1}{2}}$ be the real midpoint of $[p_i, p_j]$. Then, the GIQ-domain of p_i is composed of the four open lattice squares adjacent to p_i where the segment $]p_j, p_{j+\frac{1}{2}}[$ has been removed for each 4-neighbour p_j of p_i.

Figure 4.14(B) illustrates the union of such subsets associated with the digitisation set P of the curve C shown in Figure 4.10. It therefore forms the GIQ-domain of P. Again, P is an accurate representation of C.

Remark 4.15:

Figure 4.14(C) shows the GIQ-domain of the digitisation set of a continuous straight segment (i.e. an 8-digital straight segment according to Definition 2.2). Since chord properties in the 8-neighbourhood space were defined using the grid-

intersect quantisation, there is a clear analogy between the GIQ-domain and the visibility polygon of an 8-digital straight segment.

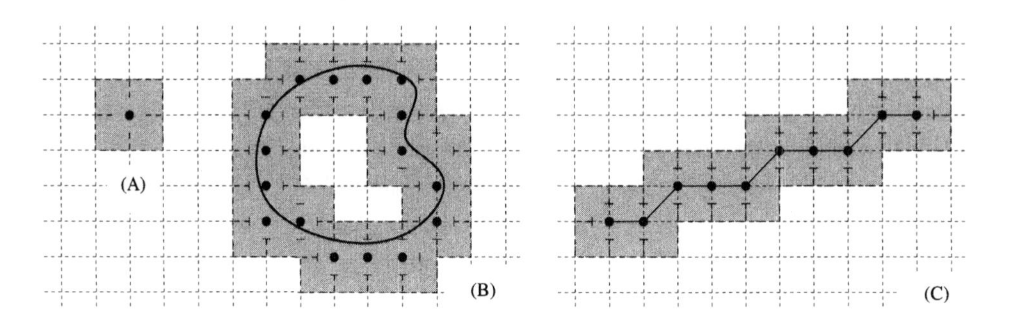

Figure 4.14 Characterisation of continuous objects associated with a digitisation set. (A) Unit continuous subset associated with each pixel p_i. (B) GIQ-domain of the digitisation set presented in Figure 4.10. (C) GIQ-domain of the digitisation set of a continuous straight segment

For comparison between GIQ- and SBQ-domains, the curve C is also digitised using the square-box quantisation. In Figure 4.15(A), black circles (•) represent such a digitisation set of C. Pixels surrounded by a square are pixels that are not present in the grid-intersect quantisation of C. Clearly, the grid-intersect quantisation reduces the number of pixels in the digitisation set. This can be understood when comparing the continuous subset shown in Figure 4.14(A) with the digitisation boxes used for the square-box quantisation.

On the other hand, Figure 4.15(B) shows the borders of the domains from the respective grid-intersect and square-box quantisations of C. The dashed line represents the border of the union of continuous subset defined by the grid-intersect quantisation of C whereas the dotted line illustrates the border of the union of digitisation boxes defined by the square-box quantisation of C.

The surface of the SBQ-domain is smaller than that of GIQ-domain. In this sense, the square-box quantisation allows for more precision in the characterisation of the preimages of a given digitisation set. However, the shape of the GIQ-domain seems to be a better descriptor of C than the shape of the SBQ-domain. This can be explained by the fact that, if the sampling step is chosen such that details in C are bigger than the size of a lattice square, they will define various intersections with lattices lines. In turn, each of these intersections is mapped onto a pixel which itself defines a continuous subset shown in Figure 4.15(A). By contrast, the square-box quantisation maps these details onto a set of 4-adjacent pixels which define a wide continuous area as result of the union of the corresponding digitisation boxes.

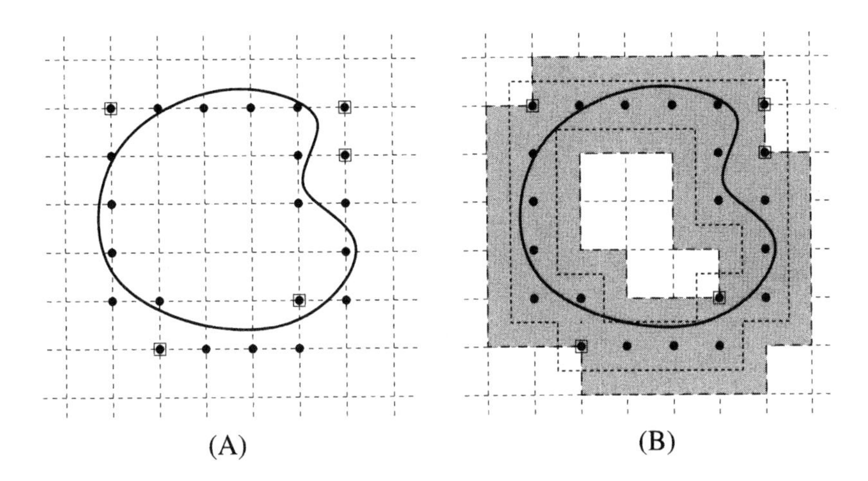

Figure 4.15 Comparison of the grid-intersect and square-box quantisation schemes. (A) Pixels added by the square-box quantisation in the digitisation set of the curve C presented in Figure 4.10. (B) SBQ- and GIQ-domains

4.1.4 Object-boundary quantisation

The object-boundary quantisation on the square lattice was introduced in [37]. This scheme is designed to map continuous sets S onto digitisation sets P that represent their borders. Typically, P is composed of the outermost pixels still belonging to S. More formally, the definition of the object-boundary quantisation of a continuous straight segment $[\alpha, \beta]$ is as follows.

Definition 4.16: Object-boundary quantisation, OBQ [37]

Given a continuous straight segment included in the line $L : y = \sigma x + \mu$ such that $0 \leq \sigma \leq 1$, $[\alpha, \beta]$, its object-boundary quantisation is composed of pixels $p_i = (x_i, y_i)$ such that $y_i = \lfloor \sigma x_i + \mu \rfloor$. The cases of other slopes are deduced by symmetry of rotation with angle of $\frac{\pi}{2}$.

Schematically, each intersection $\gamma = (x_\gamma, y_\gamma)$ between $[\alpha, \beta]$ and a vertical lattice line $(x = x_i)$ is mapped onto its closest pixel of lower y-coordinate $p_i = (x_\gamma, \lfloor \sigma x_\gamma + \mu \rfloor)$. Figure 4.16 illustrates the object-boundary quantisation process.

In Figure 4.16(A), the object-boundary quantisation is applied to the continuous set S delimited by the continuous curve C used in Figure 4.10 and results in the digitisation set shown as black discs (•). In Figure 4.16(B), this digitisation technique is applied to a continuous segment $[\alpha, \beta]$ as described in Definition 4.16 ($[\alpha, \beta]$ is also the continuous segment used in Figure 4.11(A)). The object-boundary quantisation of a continuous segment $[\alpha, \beta]$ is included in the object-boundary quantisation of the lower half-plane limited by the line L that supports $[\alpha, \beta]$ (e.g. see Figure 4.16(B)).

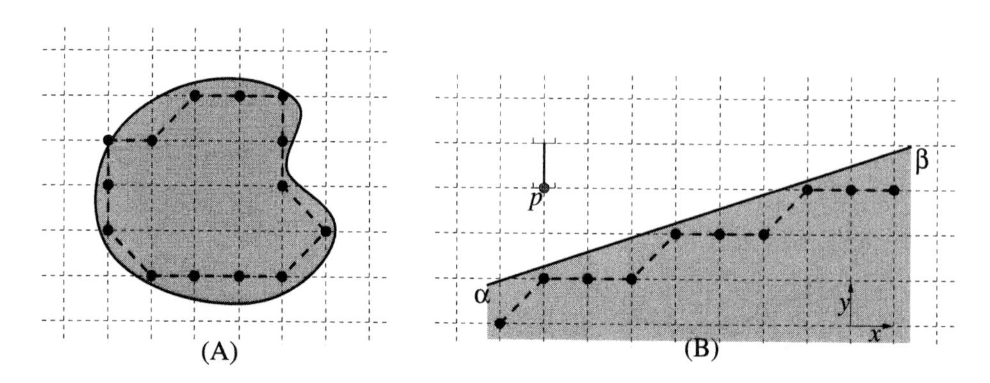

Figure 4.16 Object-boundary quantisation. (A) Continuous set S. (B) Continuous straight segment

Remark 4.17:

In the case where the border of S in not a single continuous connected set (e.g. S contains a hole S'), the object-boundary quantisation of S is formed by the object-boundary quantisation of the external border of S (i.e. the object-boundary quantisation of the set S without the hole S') and the object-boundary quantisation of S'^{c}, the complementary set of S'.

Since the object-boundary quantisation can be applied both to continuous sets and to thin curves, this digitisation technique can be compared with both the square box quantisation and the grid-intersect quantisation schemes. The object-boundary quantisation of the continuous set S is clearly included in the square-box quantisation of the same set S. Moreover, it results in a digitisation set which is 8-connected, but generally not 4-connected.

By definition, pixels in the grid-intersect quantisation of a curve C are better distributed around C than pixels in the object-boundary quantisation of the set S defined by C. The object-boundary quantisation of a continuous straight segment $[\alpha, \beta]$ whose slope σ is such that $0 \leq \sigma \leq 1$ is consistent with its grid-intersect quantisation in the sense that it is also defined by its intersections with vertical lattice lines only. Moreover, the following proposition holds.

Proposition 4.18:

The object-boundary quantisation of a continuous straight segment $[\alpha, \beta]$ having a slope σ such that $0 \leq \sigma < 1$ is equal to the grid-intersect quantisation of the continuous straight segment $[\alpha', \beta']$ with $\alpha' = (x_{\alpha'}, y_{\alpha'})$ and $\beta' = (x_{\beta'}, y_{\beta'})$ such that $x_{\alpha'} = x_{\alpha}$, $y_{\alpha'} = y_{\alpha} + \frac{1}{2}$ and $x_{\beta'} = x_{\beta}$, $y_{\beta'} = y_{\beta} + \frac{1}{2}$.

Proof:
The object-boundary quantisation of $[\alpha, \beta]$ included in $L : y = \sigma x + \mu$ where $0 \leq \sigma < 1$ is defined by the pixels $p_i = (x_i, \lfloor \sigma x_i + \mu \rfloor)$. Similarly, Remark 4.14

points out the fact that the grid-intersect quantisation of $[\alpha, \beta]$ is defined by pixels $p'_i = (x'_i, \text{round}(\sigma x'_i + \mu))$ if $0 \leq \sigma < 1$. By noting that $\text{round}(x) = \lfloor x + \frac{1}{2} \rfloor$, then, clearly, $p'_i = (x'_i, \lfloor \sigma x'_i + \mu + \frac{1}{2} \rfloor)$. Therefore, the set of pixels p'_i is also the object-boundary quantisation of a continuous straight segment $[\alpha', \beta']$ included in the line $L' : y = \sigma x + \mu + \frac{1}{2}$ and delimited by the vertical lines $(x = x_\alpha)$ and $(x = x_\beta)$. Therefore, Proposition 4.18 holds. □

Proposition 4.18 creates a relationship between the object-boundary quantisation and the grid-intersect quantisation. From then on, properties of the grid-intersect quantisation of a continuous straight segment directly map onto properties for the object-boundary quantisation of a continuous straight segment.

Proposition 4.19:

Given a continuous straight segment $[\alpha, \beta]$, let P be its object-boundary quantisation. Then,

(i) P is an 8-digital arc.

(ii) P satisfies the chord property (see Proposition 2.6). This property implies that all properties of an 8-digital straight segment are satisfied by P.

(iii) The chain-code of P can be deduced from the chain-code of P', the grid-intersect quantisation of $[\alpha, \beta]$ using the shift(.) *operator defined in Definition 2.11.*

Proof:

(i) and (ii): Follow directly from Proposition 4.18.

(iii): Let P_{pq} and $P_{p'q'}$ be the 8-digital arcs corresponding to P and P', respectively. Since P_{pq} and $P_{p'q'}$ respectively result from the grid-intersect quantisation of two parallel continuous straight segments, the continuous segments $[p, q]$ and $[p', q']$ are also parallel. Now, P_{pq} and $P_{p'q'}$ are 8-digital straight segments (see (ii)). Therefore, $P_{p'q'}$ is contained between the upper and lower bounds of P_{pq}. Hence, the chain-codes of P_{pq} and $P_{p'q'}$ can be deduced from each other by the shift(.) operator defined in Definition 2.11 (see the study of upper and lower bounds of a digital straight segment in Section 2.2.1 for more details). □

We return to our earlier Proposition 4.4, which recommends the criteria of an acceptable digitisation scheme. These criteria are investigated via examples for the object-boundary quantisation.

By definition, the object-boundary quantisation of a non-empty continuous set S is non-empty if and only if S contains at least one integer point p_i. Therefore, sets S that are totally included in a lattice square defined by four 4-connected pixels are considered as negligible. This shortcoming can be resolved by choosing a suitable sampling step h for the digitisation of S (see Definition 4.3). However, unlike the trivial digitisation scheme presented in Section 4.1.1, the definition of the object-boundary quantisation scheme guarantees that the digitisation set of a continuous straight segment $[\alpha, \beta]$ is non-empty as soon as $[\alpha, \beta]$ intersects a lattice line.

The aliasing created by the object-boundary quantisation is illustrated with the examples shown in Figure 4.17. The same sets defined by the curves that were first used in Figure 4.13 are now used as examples.

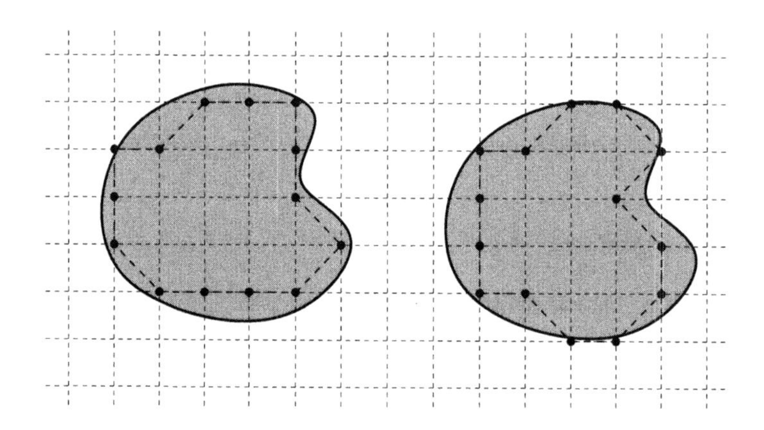

Figure 4.17 Aliasing created by the object-boundary quantisation

Clearly, since the pixels are bounded to belong to the interior of S, the aliasing created by the object-boundary quantisation is greater than that of the grid-intersect quantisation. The changes in the shape of the dashed curve representing the digitisation set of S for each position of S on the grid are clearly more important here than that shown in Figure 4.13.

Given a digitisation set P, the characterisation of the OBQ-preimages of P is, in general, not straightforward. However, in the particular case where P is a digital straight segment, the definition of the object-boundary quantisation is given by Definition 4.16, which readily suggests a trivial characterisation of the OBQ-domain of P, as shown in Figure 4.18 as a shaded area. The alternative dashed and continuous borders of this set highlight the fact that this set is semi-open.

In [37], the minimal OBQ-domain of an 8-digital straight segment is characterised by the values σ and μ (i.e. the slope and the shift) of continuous segments. These values are analytically defined using the parameters (n, q, p, s) of the 8-digital straight segment in question (see the study of parametrisation of 8-digital straight segments in Section 2.2.1).

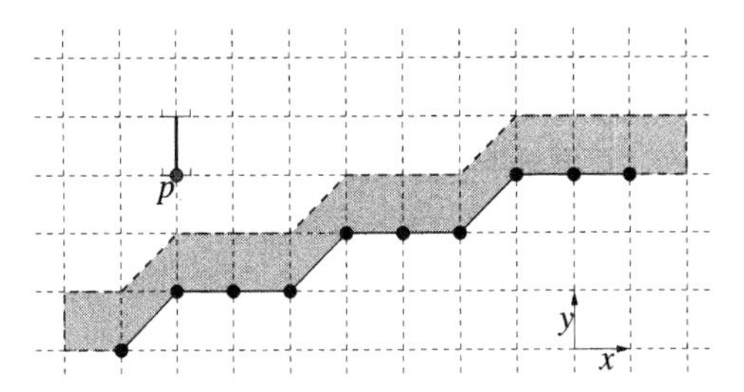

Figure 4.18 OBQ-domain of a digital straight segment

4.1.5 Summary

Three digitisation techniques have been presented, which correspond to the description of an acceptable digitisation scheme given by Proposition 4.4. Depending on whether the partition or its dual lattice is used to characterise the digitisation set, 4- or 8-connected digitisation sets are obtained, respectively (note that these two respective connectivity relationships are also dual, see Table 1.1). In each case, the digitisation scheme is designed such that geometrical properties can be studied on such discrete sets (see Chapter 2).

In particular, using the square-box quantisation as a digitisation model, cellular properties of discrete 4-connected sets can be studied. Both grid-intersect quantisation and object-boundary quantisation models lead to 8-connected digitisation sets. As described in detail in Chapter 2, characterisation of discrete straightness and discrete convexity are typically given using such sets.

It was, however, argued in Remark 4.10 that the digitisation schemes using digitisation boxes (i.e. based on the partition rather than the lattice itself) model the real acquisition process more accurately. In the case of binary images, a physical captor associated with each pixel area would output a binary (0-1) value, depending on whether the continuous set S that is to be digitised intersects this area or not. Alternative models exist where the output binary value depends on the portion of S present in each digitisation box B_i. More formally, if $f(x, y)$ is the two-dimensional binary function representing the input binary image, the pixel $p_i = (x_i, y_i)$ corresponding to the digitisation box B_i is considered as black if and only if

$$\frac{1}{|B_i|} \iint_{B_i} f(x, y)\, dx\, dy > \tau$$

where $\tau \in [0, 1]$ is a given threshold and $|B_i|$ the area of B_i. However, using such schemes, some properties such as connectivity in the digitisation sets may not be preserved. Further details on such models can be found, for example, in [116].

4.2 Storage of binary digital images

The output of the digitisation process detailed in Section 4.1 is typically a set of discrete points in which each point is associated with a binary value which represents the black or white colour of the corresponding pixel (see also Section 1.1).

In this section, storage of binary digital images is considered. Different criteria define an efficient storage. These criteria are readily given by the context in which this storage will be used. Typically, accessibility time and storage volume are two such criteria. The respective importance of these two parameters varies upon the type of application developed. Firstly, depending on whether the application (e.g. image recognition, image enhancement) is to operate on-line (i.e. in real-time) or off-line (i.e. in a batch process), accessibility time becomes crucial and storage volume constraints can be relaxed in the first case, whereas the opposite situation arises in the second case.

Moreover, accessibility time greatly depends on the approach taken in the development of the theoretical model. In this section, we consider the case of the classic pixel-by-pixel approach for which a matrix-type storage is suitable. Additionally, we introduce specialised structures which are relevant when using the graph-theoretic approach presented in Chapter 3 and developed throughout this book. Section 4.2.1 addresses the problem of defining the storage of a binary digital image that is suitable for a given processing.

If the image is to be stored in a database-type environment, storage volume clearly becomes a crucial issue. Data compression typically reduces the storage volume with the cost of an increased accessibility time. In this sense, Section 4.2.2 presents a brief introduction to basic techniques used for binary data compression and also provides references from which the interested reader can find further details on these techniques.

4.2.1 Storage

4.2.1.1 Matrix storage

A binary digital image I is formally represented by a set of integer points $p_i = (x_i, y_i)$ associated with binary (0-1) values. By convention, an integer point associated with a value of 1 (respectively 0) represents a black pixel (respectively white pixel) in the image. Moreover, it is mostly the case that such integer points form a rectangular area of size $W \times H$, where W and H are the width and the height (in pixels) of the image, respectively. Therefore, a trivial method for storing I is used a matrix $\mathcal{M}(I) = (m_{kl})_{k=0,\ldots,W-1;l=0,\ldots,H-1}$ of 0's and 1's. $\mathcal{M}(I)$ is defined such that m_{kl} is the binary value associated with pixel $p_i = (x_i, y_i)$ with $x_i = k$ and $y_i = H - l - 1$.

Remark 4.20:

The fact that $y_i = H - l - 1$ *is simply due to the fact that the vertical axis is classically oriented in opposite directions in the plane and in a matrix (i.e. upwards in the plane and downwards in a matrix).*

Example: Matrix storage of a binary digital image

Consider the binary digital image I given in Figure 4.19(A). Black and white pixels are represented by black and empty discs, respectively. Clearly the size of I is $W = 5$ and $H = 6$. The matrix $\mathcal{M}(I)$ that represents I is given in Figure 4.19(B).

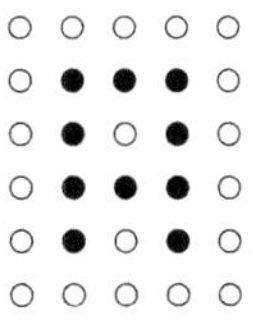

○ ○ ○ ○ ○
○ ● ● ● ○
○ ● ○ ● ○
○ ● ● ● ○
○ ● ○ ● ○
○ ○ ○ ○ ○

(A)

$$\mathcal{M}(I) = \begin{bmatrix} 0 & 0 & 0 & 0 & 0 \\ 0 & 1 & 1 & 1 & 0 \\ 0 & 1 & 0 & 1 & 0 \\ 0 & 1 & 1 & 1 & 0 \\ 0 & 1 & 0 & 1 & 0 \\ 0 & 0 & 0 & 0 & 0 \end{bmatrix}$$

(B)

Figure 4.19 Example of a binary digital image

Matrix storage is well-suited for operations that involve a pixel and its complete neighbourhood (i.e. black and white pixels). This type of processing includes operators based on masks such as distance transformations and operators derived from mathematical morphology [149, 150] (e.g. see Chapter 5). The colour (i.e. black or white) of a pixel can be directly stored in the matrix. However, matrix-storage involves white pixels and it is often the case that the number of black pixels is much smaller than that of white pixels in a binary digital image. In this context, a storage involving black pixels only would therefore be more efficient.

4.2.1.2 Graph-theoretic context

The graph-theoretic approach presented throughout this book is essentially based on black pixels in the image, since it is concerned with the processing of the foreground only. In this respect, we present storages of binary digital images which directly follow from algorithmic graph theory as presented in Chapter 3 and allow for the storage of the foreground of the image only. Static and dynamic graph-theoretical storages exist which can be used depending on whether the image is to evolve during the process or not. The static and dynamic forward star structure are presented here as examples of such storage methods, respectively.

In Section 3.3 the analogy between an image I and a graph G was detailed. The grid-graph of a subset of pixels in I (e.g. the foreground) allows for the representation of both the pixels and their neighbourhood relationships.

The storage of such graph is detailed here. Firstly, the static forward star structure is presented. This structure efficiently stores sparse graphs, such as grid graphs (see Proposition 3.22) and has a wide number of applications [34, 49].

Definition 4.21: Static forward star structure

Given a graph $G = (V, A)$, G can be stored in the forward star structure defined as follows. Two arrays $\mathcal{S}$ and $\mathcal{T}$ composed of elements $\mathcal{S}_l$ and $\mathcal{T}_l$, respectively, are needed. $\mathcal{S}$ is of size $(N_G + 1)$ and $\mathcal{T}$ is of size M_G if G is a directed graph and $(2 \times M_G)$ if G is undirected. For each vertex $u_i \in V$, $\mathcal{T}$ contains the list of successors of u_i starting from the entry numbered $\mathcal{S}_i$. Clearly, if G is an undirected graph, each arc (u_i, u_j) is stored twice as (u_i, u_j) and (u_j, u_i).

Example: Forward star structure for the storage of a binary digital image

For clarity, the image shown in Figure 4.20(A) is used as example and the 8-neighbourhood relationship is considered. Figure 4.20(B) displays the corresponding (undirected) 8-grid graph of this image. Finally, Figure 4.20(C) illustrates the forward star structure that represents this graph.

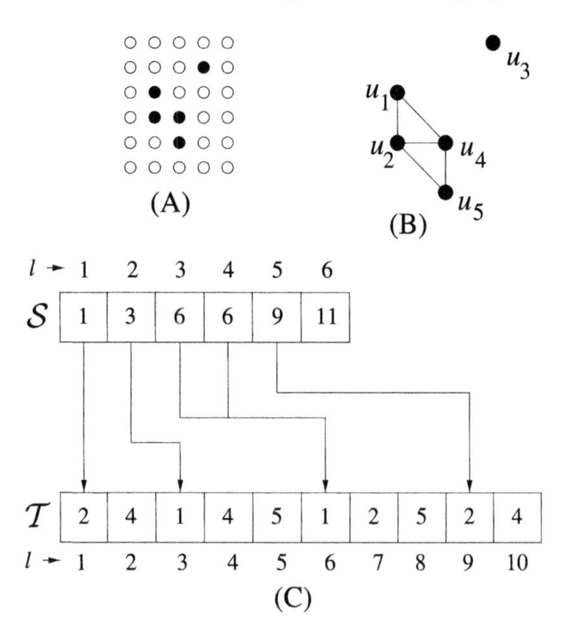

Figure 4.20 Forward star structure of a grid graph. (A) Binary digital image. (B) Grid graph. (C) Forward star structure

Clearly, $\mathcal{T}$ is of size $(2 \times M_G) = 10$. The indices of the successors of a vertex u_i in $\mathcal{S}$ are listed from $\mathcal{T}_{(\mathcal{S}_i)}$ to $\mathcal{T}_{(\mathcal{S}_{i+1}-1)}$, inclusive. This explains why $\mathcal{S}$ contains an extra value. The particular case of u_3, which does not have any neighbour, is marked by the values $\mathcal{S}_3 = 6 > \mathcal{S}_4 - 1 = 5$.

Two main advantages of the forward star structure can be distinguished. Firstly, since it is specialised for the storage of sparse graphs, information retrieval is direct and very efficient [34]. Moreover, this structure allows for the storage of black pixels only. Such a characteristic makes this storage independent of the context in which the image is used (e.g. the image can equivalently be stand-alone or a component of a main image). By contrast, a preprocessing step which removes all redundant horizontal and vertical white border lines is necessary for the matrix storage to be efficient. Moreover, this structure can clearly be constructed in $\mathcal{O}(W \times H)$ operations from the matrix storage since all black pixels in the forward star of a given pixel can be characterised in constant time from the matrix storage.

In a process where the image is to be modified, the grid graph of the image is also to be updated by the addition or removal of vertices. This is the case, for example, when the colour of some pixels is to be switched from black to white or vice versa in a smoothing or noise reduction process. The grid graph associated with black pixels only is therefore changing throughout the process. In this context, the static forward star structure is not an efficient representation for the grid graph. The dynamic forward star structure which relies on two mutually exclusive numbering systems for the vertices in a graph allows for the dynamic storage of such graphs [26].

Definition 4.22: Dynamic forward star structure [26]

Given a graph $G = (V, A)$, the dynamic forward star structure of G is composed of two pairs of arrays $(\mathcal{S}^k, \mathcal{T}^k)$, $k = 1, 2$ of elements $\mathcal{S}_l^k$, $\mathcal{T}_l^k$, respectively. Arrays $\mathcal{S}^k$ are of size N_G and arrays $\mathcal{T}^k$ are of size M_G. Both elements $\mathcal{S}_l^k$ and $\mathcal{T}_l^k$ are references to either an element $\mathcal{S}_l^k$ or an element $\mathcal{T}_l^k$. The reference to an element being made by storing its index. By convention, a positive reference l refers to an element $\mathcal{T}_l^k$ and a negative reference $(-l)$ to an element $\mathcal{S}_l^k$.

Let us assume that all arcs $(u_i, u_j) \in A$ are stored sequentially (in any order) in an array $\mathcal{A}$ of elements $\mathcal{A}_l = (u_i, u_j)$. With reference to $\mathcal{A}$, for each vertex $u_i \in V$, $\mathcal{S}_i^1$ points to the element $\mathcal{T}_l^1$ (i.e. contains the reference l) such that $\mathcal{A}_l$ is the first arc of the form (u_i, u_j) such that $j > i$. If no such arc exists, $\mathcal{S}_i^1$ points to itself (i.e. contains the reference $-i$). Similarly, for each vertex $u_i \in V$, $\mathcal{S}_i^2$ points to the element $\mathcal{T}_l^2$ (i.e. contains the reference l) such that $\mathcal{A}_l$ is the first arc of the form $(u_i, u_j) \in A$ such that $j < i$. If no such arc exists, $\mathcal{S}_i^2$ points to itself (i.e. contains the reference $-i$).

Given the arc $(u_i, u_j) = \mathcal{A}_l$, an element $\mathcal{T}_l^1$ points to another element $\mathcal{T}_{l'}^1$, where $\mathcal{A}_{l'}$ is the subsequent arc of the form $(u_i, u_{j'})$ where $j' > i$. If no such arc exists, $\mathcal{T}_i^1$ points to $\mathcal{S}_i^1$ (i.e. contains the reference $-i$). Similarly, given the arc $(u_i, u_j) = \mathcal{A}_l$, an element $\mathcal{T}_l^2$ is a reference to another element $\mathcal{T}_{l'}^2$, where $\mathcal{A}_{l'}$ is the the subsequent arc of the form $(u_i, u_{j'})$ where $j' < i$. If no such arc *exists, $\mathcal{T}_i^2$ points to $\mathcal{S}_i^2$ (i.e. contains the reference $-i$). Each pair of arrays $(\mathcal{S}^k, \mathcal{T}^k)$, $k = 1, 2$ therefore forms a circular list which starts and ends at each vertex $u_i \in V$. It can easily be shown that, using such a structure, the array $\mathcal{A}$ becomes redundant and only arrays $\mathcal{S}^k$ and $\mathcal{T}^k$, $k = 1, 2$ are needed to describe completely the graph $G = (V, A)$.*

Example: Dynamic forward star structure for the storage of a binary digital image

The graph $G = (V, A)$ used in this example is presented in Figure 4.20(B). The dynamic forward star structure of this graph is presented in Figure 4.21. Firstly, the array $\mathcal{A}$ that contains all arcs in A is represented as dotted lines since it is not truly part of the dynamic forward star structure. Arrays $\mathcal{S}^1$, $\mathcal{T}^1$ and $\mathcal{S}^2$, $\mathcal{T}^2$ are related to each other by arrows representing pointers.

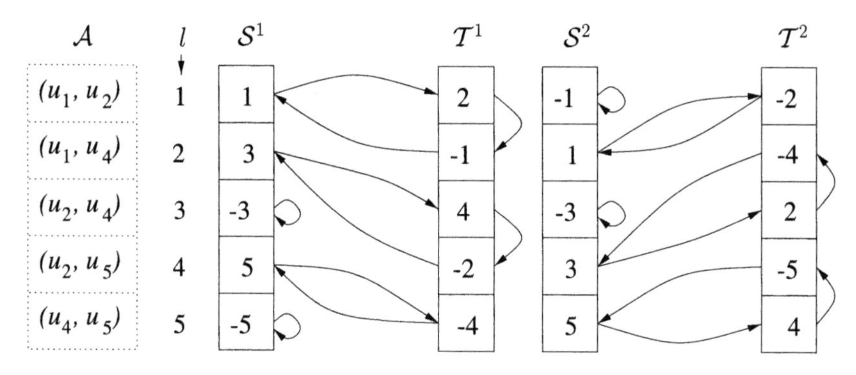

Figure 4.21 Dynamic forward star structure of the grid graph shown in Figure 4.20(B)

Since G is undirected, $(u_i, u_j) = (u_j, u_i)$. The array $\mathcal{A}$ is a sequential list of undirected arcs stored in any order. The construction of the dynamic forward star structure of G is illustrated for the case of vertex u_4. By definition, $\mathcal{S}_4^1$ points to $\mathcal{T}_5^1$, where $l = 5$ is the index in $\mathcal{A}$ of the first arc of the form (u_4, u_j) such that $j > 4$ (i.e. the arc $\mathcal{A}_5 = (u_4, u_5)$). Therefore, $\mathcal{S}_4^1 = 5$. Since there is no further arc (u_4, u_j) such that $j > 4$, $\mathcal{T}_5^1$ points back to $\mathcal{S}_4^1$ and therefore $\mathcal{T}_5^1 = -4$. Similarly, $\mathcal{S}_4^2$ points to $\mathcal{T}_3^2$ where $l = 3$ is the index in $\mathcal{A}$ of the first arc of the form (u_4, u_j) such that $j < 4$ (i.e. the arc $\mathcal{A}_5 = (u_4, u_2) = (u_2, u_4)$, since G is undirected). Therefore, $\mathcal{S}_4^2 = 3$. Now, by definition, $\mathcal{T}_3^2$ point to $\mathcal{T}_l^2$, where $\mathcal{A}_l$ is the next arc of the form (u_4, u_j) such that $j < 4$. This arc is $\mathcal{A}_2 = (u_4, u_1) = (u_1, u_4)$ (i.e. $l = 2$). Therefore $\mathcal{T}_3^2 = 2$. Finally, since there are no more such arcs in $\mathcal{A}$, $\mathcal{T}_2^2$ points to $\mathcal{S}_4^2$ (i.e. $\mathcal{T}_2^2 - 4$).

Clearly, $\mathcal{S}_1^2$ points to itself since there exists no arc (u_1, u_j) such that $j < 1$. Similarly, $\mathcal{S}_5^1$ points to itself. The special case of u_3 whose forward star is empty is marked by the fact that $\mathcal{S}_3^1 = \mathcal{S}_3^2 = -3$ (i.e. both $\mathcal{S}_3^1$ and $\mathcal{S}_3^2$ point to themselves). In general, if u_i is disconnected (i.e. forms a connected component in itself), $\mathcal{S}_i^1 = \mathcal{S}_i^2 = -i$.

Clearly, the array $\mathcal{A}$ is not needed for the characterisation of the forward star of a vertex u_1. Using the previous example, the procedure for characterising the forward star of vertex u_2 is as follows. Starting from $\mathcal{S}_2^k$, both circular lists are investigated ($k = 1, 2$) and arc indices in this list are stored. Starting from $\mathcal{S}_2^1$, the first circular list passes through $\mathcal{T}_3^1$ and $\mathcal{T}_4^1$. This means that arcs $\mathcal{A}_3$ and $\mathcal{A}_4$ are of the form (u_2, u_j) where $j > 2$. Following the circular lists from $\mathcal{T}_3^2$ and $\mathcal{T}_4^2$, they end in $\mathcal{S}_4^2$ and $\mathcal{S}_5^2$, respectively. Therefore, $\mathcal{A}_3 = (u_2, u_4)$ and $\mathcal{A}_4 = (u_2, u_5)$. Using a similar procedure (and using the other set of circular lists), one finds that $\mathcal{A}_1$ is an arc of the form (u_2, u_j) where $j < 2$. Starting from $\mathcal{T}^1$, the circular list ends in $\mathcal{S}_1^1$ and, therefore, $\mathcal{A}_1 = (u_2, u_1)$. Therefore, the forward star of vertex u_2 is $\{u_1, u_4, u_5\}$.

The fact that arrays $\mathcal{T}^k$ do not depend on the order in which the arcs appear suggests that the insertion or the removal of an arc can be done efficiently. The addition or the removal of a vertex can also be done efficiently if the arrays $\mathcal{S}^k$ are stored as chained-lists. Such procedures are detailed in [26].

4.2.2 Binary data compression

Our objective here is to give the reader a simple introduction to binary data compression. Techniques presented here often form the basis for the development of more specialised or more efficient compression techniques. Further developments concerning binary image compression will also be presented throughout Chapter 7 For details on general image compression techniques, the interested reader is referred to [19, 50, 109].

The aim of data compression is to reduce the size of the storage required when constraints on accessibility of data are relaxed (e.g. for mass storage). Four basic techniques for achieving binary data compression are summarised here. In all of the techniques, pixels of the same colour are grouped in segments or blocks and a code is associated with each of these blocks that allows for information retrieval. Techniques presented here allow for compression of data without any loss of information (i.e. all information contained in the data is stored).

Binary data (i.e. black and white pixels) can be represented by a set of 0's and 1's. In this context, the storage unit is chosen as a bit which can take these two values.

4.2.2.1 White block skipping (WBS)

It is generally the case that, in a binary digital image, the number of white pixels is much larger than the number of black pixels. It therefore seems natural that skipping white pixel blocks would allow for efficiently reducing the size of the storage. If the matrix that stores the image is read line by line, a one-dimensional (1D) signal may be obtained as a succession of 0's and 1's. One-dimensional white block skipping technique operates as follows to represent such a signal.

Definition 4.23: One-dimensional white block skipping (1D-WBS)

Given an integer $n > 0$ as segment size, the signal is decomposed into n-pixel segments. A segment which contains only white pixels (0's) is coded with one bit set to '0' and a segment which contains at least one black pixel ('1') is coded with the $(n+1)$ bit of the original segment data prefixed by a bit set to '1'.

Example: 1D-WBS

Consider the image represented in Figure 4.22(A). Setting $n = 4$, the one-dimensional partition of the image is shown in Figure 4.22(B) and yields the original one-dimensional signal:

0000 0001 1000 0110 0000 0100 0000 1000 0000

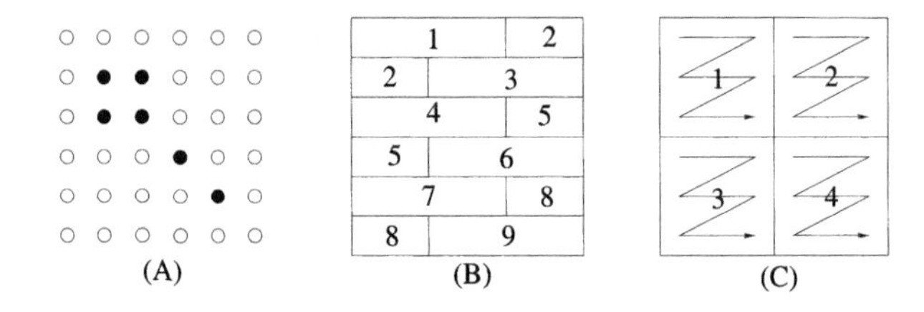

Figure 4.22 Test image for the white block skipping coding scheme. (A) Original image. (B) Partition used for the 1D-WBS. (C) Partition used for the 2D-WBS

Using 1D-WBS, the resulting coded signal is

0 10001 11000 10110 0 10100 10000 11000 0

A two-dimensional (2D) structure of the data is readily defined in the binary image. A two-dimensional white block skipping technique which takes advantage of this specific form of the data is a simple extension of 1D-WBS described above.

Definition 4.24: Two-dimensional white block skipping (2D-WBS)

Given an integer $n > 0$ as block size, the signal is decomposed in $(n \times n)$-pixels blocks. A block which contains only white pixels (0's) is coded with one bit set to '0' and a segment which contains at least one black pixel ('1') is coded with the $(n^2 + 1)$ bits of the original block data prefixed by a bit set to '1'.

Example: 2D-WBS

Consider again the image represented in Figure 4.22(A). Setting $n = 3$, the image partition is as shown in Figure 4.22(C). Arrows represent the way pixels are traversed in each block. Using 2D-WBS, the resulting coded signal is then

$$1000011011\ 0\ 0\ 1100010000$$

Information retrieval for both codes is then simply done as follows. Given n (which can be defined as standard or added in the code), the first bit is read. If it equals '0', n (for 1D-WBS) or n^2 (for 2D-WBS), white pixels are written in the output. Else, it equals '1' and the following n (for 1D-WBS) or n^2 (for 2D-WBS) bit are read and written as pixels in the output. The next block can then be processed in a similar way.

Clearly, white block skipping coding schemes are efficient in images where black pixels can be grouped into a small number of connected components. In this case white segments or blocks that separate the connected components of the foreground can be found. In the case where only small white segments or blocks can be found, WBS coding may increase the size of the storage.

Similarly, these schemes are highly sensitive to noise since the addition of one black pixel in a white segment (respectively, block) adds n (respectively, $n \times n$) pixels to the resulting coded signal.

4.2.2.2 Run length encoding (RLE)

This one-dimensional coding scheme is designed to be efficient on signals where black and white pixels appear in runs (i.e. where there is a small number of isolated black or white pixels in the signal). In this case, there is no restriction on how the numbers of black and white pixels compare in the image.

Definition 4.25: Run length encoding (RLE)

Given a one-dimensional signal as a succession of 0's and 1's, assuming that the first pixel is white ('0'), the signal is traversed and lengths of runs formed by pixels of identical colour are successively added to the code.

Example: Run length encoding

Consider again Figure 4.22(A). The corresponding one-dimensional signal

$$00000001100001100000010000001000000 0$$

is coded as

$$7, 2, 4, 2, 6, 1, 6, 1, 7$$

Information retrieval is then done as follows. Given the colour of the first pixel (which can be added in the code), the first run length is read and the appropriate number of pixels of the corresponding colour are written in the output. The colour is then inverted and another run-length is read and so on.

Run lengths are then to be themselves encoded in a suitable form (e.g. binary coding). Since the range of values a run length can take is large (i.e. from 1 to the size of the image), codes of variable length have been designed to encode the run lengths. Typically, a code is associated with each possible run length depending on the frequency of its occurrence. A run length that occurs frequently will be associated with a short code, whereas a longer code will be associated with a run length that rarely appears. Huffman codes are known to be the most efficient such codes but are often found hard to implement. Other suboptimal techniques exist that are easier to implement. These coding techniques are outside the scope of this book and are not detailed here. For an extended analysis, the interested reader is invited to consult [109].

4.2.2.3 Quadtrees

This coding scheme relies on a recursive decomposition of the image. This decomposition yields a hierarchical structure (i.e. the quadtree) that represents the image. A simple way of representing quadtrees is presented here. We introduce quadtree based coding techniques via the following example.

Example: Coding a binary digital image using a quadtree

Consider the image shown in Figure 4.23(A). At each stage of the decomposition, each block is decomposed into four blocks of equal size. Whenever a block contains pixels of one colour (black or white) only, it is not decomposed further (see Figure 4.23(B)).

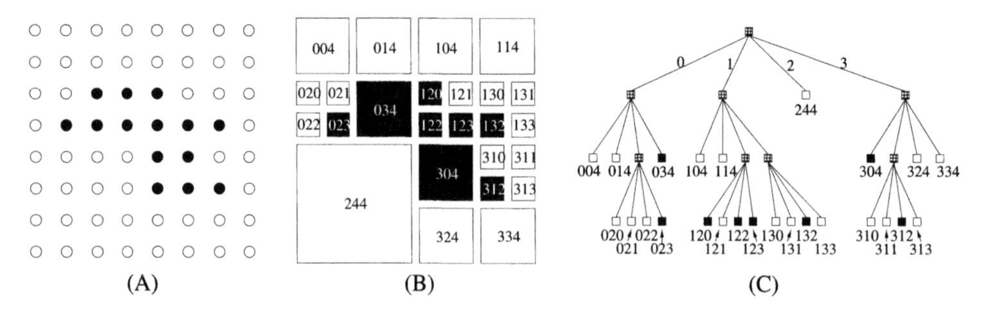

Figure 4.23 Test image for the quadtree encoding scheme. (A) Original image. (B) Recursive decomposition. (C) Quadtree and corresponding codes

This decomposition can then be represented using the quadtree shown in Figure 4.23(C), where a node corresponds to a block and an arc is read as 'is decomposed into'. The root of the quadtree represents the complete image and a leaf represents a block of uniform (black or white) colour. At each level nodes are ordered upon the block they correspond to, following the sequence NW, NE, SW, SE.

Each of the leaves of the quadtree is then associated with a code as follows. At each level, blocks are numbered from 0 to 3 with respect to their position (NW, NE, SW or SE) in the parent block. In Figure 4.23(C), such a numbering scheme is illustrated for the upper-most level in the quadtree. A leaf corresponding to a uniform block is associated with the sequence of values met during the traversal of the quadtree from the root to this leaf. For each leaf, a code of constant length is to be determined. The length of this code is given by the maximal level of decomposition in the image. In this respect, a shorter coding sequence is completed by appending 4's at its end. These codes are shown in Figure 4.23(B), superimposed on the block they correspond to and in Figure 4.23(C), associated with their respective leaves.

Clearly, the image can be characterised by black blocks only. Therefore, the image can be uniquely coded by storing the black leaves of the quadtree only. Following this, the image shown in Figure 4.23(A) is represented by the code sequence

$$023\ 034\ 120\ 122\ 123\ 132\ 304\ 312$$

Leaf codes can then simply be encoded as 0's and 1's by a classic fixed length binary coding technique (e.g. with 9 bits per leaf in this case).

Image retrieval is done as follows. Given the maximal level of decomposition in the image, leaf codes are successively read and the block they correspond to are added as black pixels to an originally white image.

For an image to be suitable for such a encoding technique, its size should be of the form $2^n \times 2^n$. More advanced decomposition schemes exist where a block may not be square. For more details on quadtrees, related structures and operations that can be performed on them, the interested reader is invited to consult [145–147].

4.2.2.4 Boundary encoding

This coding scheme was introduced and extensively studied in [43]. Typically, it is designed for images in which the foreground is composed of a small number of connected components. Each connected component is completely defined by the chain-code of its boundary and a starting point (see Definition 2.3).

Definition 4.26: Boundary encoding

A complete binary image is characterised by a set of pairs (p_j, C_j) *where* $p_j = (x_j, y_j)$ *is the starting point of the chain-code* C_j *of the boundary of the object* j *in the image.*

Example: Boundary encoding

Consider the image shown in Figure 4.24(A).

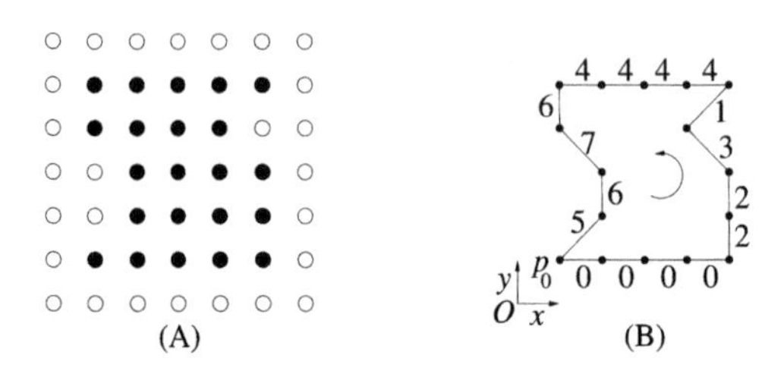

Figure 4.24 Test image for the boundary encoding scheme. (A) Original image. (B) Boundary chain-code

As shown in Figure 4.24(B), starting from $p_0 = (1,1)$ with the counter-clockwise orientation, the chain-code of the boundary of the object contained in this image is

$$\{0,0,0,0,2,2,3,1,4,4,4,4,6,7,6,5\}$$

The origin O is chosen as the bottom left corner of the image. This can be defined by convention. Therefore, this image can be coded as

$$110000223144446765$$

Using such a coding sequence, by first reconstructing boundaries of connected components in the foreground and then using a filling technique, the set of all black pixels can be retrieved, giving the original image.

Compression techniques presented in this chapter are suitable for any one- or two-dimensional binary digital signal and process the signal at the pixel level. They allow for efficient data storage while conserving the complete information. Other techniques exist where the information is simplified before or during storage (e.g. smoothing). The binary image restored from such a storage will therefore include some differences from the original image, which are assumed to be minor. Higher level compression techniques (called representations, in this case) will be presented when studying image representations in Chapter 7.

4.2.2.5 JBIG

JBIG (Joint Bi-level Expert Group) refers to an international research group whose aim is to design a compression technique for binary images, in order to improve compression of facsimile images. This development follows the line of the work done by the Joint Photographic Expert Group (JPEG) whose compression interchange format for colour images is now widely used.

The core of the JBIG compression technique is an arithmetic coder that encodes binary pixel values. This coder is the same as the one used in the JPEG standard. The encoding of the binary value at pixel p is done using a group of pixels which are neighbours to p (known as the "neighbouring pixel template") to encode the statistics related to p. An example of such a neighbouring pixel template is shown in Figure 4.25, where the values at pixels are used to "predict" the value at pixel p. Since the encoding is done in a row-scanning order, this template indicates causality and only considers already encoded pixel values.

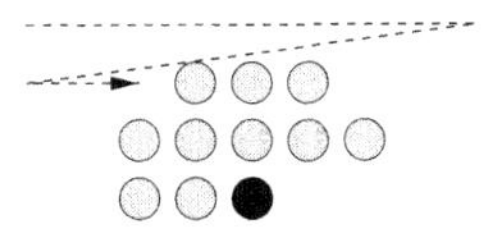

Figure 4.25 Neighbouring pixel template for JBIG encoding. The arrow indicates the row-scanning order and the black dot (•) is the pixel to be encoded

The JBIG compression technique has been designed for and proved to be particularly efficient on the class of half-tone pictures (binary approximations of grey-level images). Since this type of binary image is not truly related to the scope of this book, the reader is referred to [117] for technical details and performance characterisation of the JBIG compression technique.

Chapter 5

DISTANCE TRANSFORMATIONS

The notion of distance is fundamental in any analysis process. Models for image representation generally involve minimum distances between pixels and borders of components, pixels closer to a given set of pixels than to another set and so on. In this context, it is often the case that measurements are operated between a point and a set rather than between points. Such an operation defines a important tool in digital image processing, the distance transformation (DT). By storing its result into the distance map (DM), the computational effort can be reduced and global properties of the image can be characterised. The operation of calculating the distance map of an image is referred to as *distance mapping*.

Distance mapping often forms the preprocessing step in an image analysis process. In most cases, following distance mapping, the image is represented by its distance map for further processing. For example, operations such as smoothing, merging and thinning rely on the computation of the distance map of the image. In turn, the distance map itself depends on the definition of the distance used for computing it. Both discrete and Euclidean distances may be used for defining discrete or Euclidean distance maps, respectively. In Chapter 1, discrete distances were presented and their compatibility with the Euclidean distance was studied. The compatibility between discrete distance maps and Euclidean distance maps mostly derives from the work presented there.

In this chapter, distance transformations and distance maps are studied. In Section 5.1, both concepts are introduced in a general context, independent from the definition of the distance used. Then, Section 5.2 analyses distance mapping in conjunction with discrete distances. Section 5.3 introduces the problem of Euclidean distance map generation and proposes algorithms for solving it. Finally, comments on these results are given in Section 5.4

5.1 Definitions and properties

Distance mapping is first introduced in a generic context. The basic operation for such a mapping is the distance transformation. Typically, the distance transformation of a set P associates with each point $p \in P$ the minimal distance value between p and any point of Γ, a subset of P (e.g. its border). More formally, the following definition is given.

Definition 5.1: Distance transformation (DT)

Given a set P and a subset Γ, given a distance function $d(.,.)$ that satisfies the metric conditions given in Definition 1.15, the distance transformation $\mathrm{DT}(.)$ *of P associates with a point $p \in P$ the value*

$$\mathrm{DT}(p) = \min_{q \in \Gamma} d(p, q)$$

The result of the distance transformation depends on the distance considered. Examples of distance transformations will be given throughout this chapter of cases where $d(.,.)$ is a discrete distance (Section 5.2) and the Euclidean distance (Section 5.3). The distance transformation of a set P results in a set of values associated with each point of P.

Definition 5.2: Distance map of a set P

The distance map of P is a matrix $(\mathrm{DT}(p))_p$ which has the same size as the original image and stores the values $\mathrm{DT}(p)$ of the distance transformation for any point $p \in P$.

Remark 5.3:

The similarity between the matrix storage of a digital image and distance maps will become apparent during this chapter.

Some properties of the distance transformation can readily be given, which do not depend on the distance used for defining it.

Proposition 5.4:

Given a set P and a subset Γ, the distance transformation of P satisfies the following properties

(i) $\mathrm{DT}(p) = 0$ *if and only if $p \in \Gamma$.*

(ii) Defining the point $\mathrm{ref}(p) \in \Gamma$ such that $\mathrm{DT}(p)$ is defined by $d(p, \mathrm{ref}(p))$ (i.e. $\mathrm{ref}(p)$ is the closest point to p in Γ), the set of points $\{p$ such that $\mathrm{ref}(p) = q\}$ forms the Voronoi cell of point q. If $\mathrm{ref}(p)$ is non-unique, then p is on the border of a Voronoi cell.

Proof:

(i): Immediate.

(ii): By definition, the Voronoi cell corresponding to a point $q \in \Gamma$ with respect to the points in Γ contains all points in P closer to q than to any other point in Γ. Moreover, the points on the border of a Voronoi cell are equidistant to at least two points in Γ. Now, by definition of the distance transformation, all points p such that $\mathrm{ref}(p) = q$ are such that $d(p,q) = \min\{d(p,p') \; ; \; p' \in \Gamma\}$. Therefore, (ii) clearly holds. □

Moreover, if Γ is the border of the set P, the following proposition holds.

Proposition 5.5:

Given a set P and its border Γ, the distance transformation of P has the following properties

(i) By definition, $\mathrm{DT}(p)$ *is the radius of the largest disc centred at p and totally included in P.*

(ii) If there is exactly one point $q \in \Gamma$ such that $\mathrm{DT}(p) = d(p,q)$, *then there exists a point $r \in P$ such that the disc of radius* $\mathrm{DT}(r)$ *centred at r totally contains the disc of radius* $\mathrm{DT}(p)$ *centred at p.*

(iii) Conversely, if there are at least two points q and q' in Γ such that $\mathrm{DT}(p) = d(p,q) = d(p,q')$, *then there is no disc totally included in P that totally includes the disc of radius* $\mathrm{DT}(p)$ *centred at p. In that case, p is said to be a* centre of maximal disc.

Proof:

(i): Immediate by definition of $\mathrm{DT}(p)$.

(ii): Consider a point p such that there exists exactly one point $q \in \Gamma$ such that $\mathrm{DT}(p) = d(p,q)$. The point $q' \in \Gamma$ is defined as an intersection between the line defined by p and q and Γ (see Figure 5.1(A)). Clearly, there exists a point $r \in [q,q']$ such that $\mathrm{DT}(p) < \mathrm{DT}(r)$ and such that $\mathrm{DT}(r) = d(r,q)$. By simple geometric considerations, it is clear that the disc of radius $\mathrm{DT}(r)$ centred at r totally contains the disc of radius $\mathrm{DT}(p)$ centred at p. The border of such a disc is represented as a dashed circle in Figure 5.1(A)). Therefore (ii) holds.

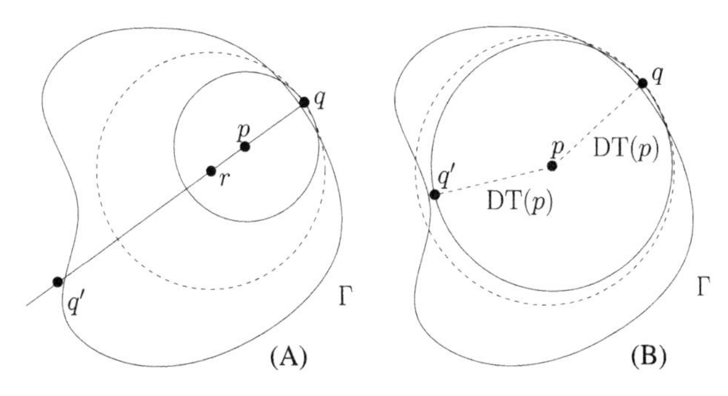

Figure 5.1 (A) p is not a centre of maximal disc. (B) p is a centre of maximal disc

(iii): Assume now that there exist two points $q \in \Gamma$ and $q' \in \Gamma$ such that $\mathrm{DT}(p) = d(p,q) = d(p,q')$. Such a configuration is illustrated in Figure 5.1(B).

It is easy to show that, in this case, any disc that contains the disc of radius DT(p) centred at p is not totally included in the set P (e.g. the disc whose border is shown as a dashed circle in Figure 5.1(B)). This is in contradiction with Proposition 5.4(ii) and, therefore, (iii) holds. □

These properties will be revisited when developing algorithms for the solution of the distance mapping problem and when detailing the relationship between distance transformations and particular structures found in computational geometry (see Section 5.4). These properties will prove useful for the development of models for image representations and will be fully exploited by algorithms presented in later chapters.

Remark 5.6:

In binary digital image processing, P represents the image, or a part of it (e.g. the foreground), and is composed of discrete points (i.e. the pixels). For such a set the definition of the border is given in Definition 1.11. In this case, the distance map can be represented by a grey-level image where the grey-level at a pixel represents the value of the distance transformation of the image at this pixel (see Section 5.4).

5.2 Discrete distance transformations

The case where the distance $d(.,.)$ used in Definition 5.1 is a discrete distance (as defined in Section 1.4) is studied in this section. Based on the conclusions derived in Section 1.4.4, we will mostly concentrate on discrete distances defined on square lattices. Extensions to hexagonal partitions (i.e. triangular lattices) will also be developed in this section.

Discrete distances and their properties were studied in Sections 1.4 and 1.5. Clearly, using Definition 5.1, any discrete distance may be used for the discrete distance mapping [111]. However, chamfer distances are typically simple to compute and have been shown to be more accurate than other discrete distances, such as hexagonal and octagonal distances, in the approximation of the Euclidean distance. The generation of discrete distance maps will therefore mostly be detailed using chamfer distances. Such discrete distance transformations and distance maps are also referred to as chamfer distance transformations and chamfer distance maps, respectively. Note that, in Remark 1.37, it was pointed out that the d_4 and d_8 distances can be seen as particular cases of chamfer distances.

Two approaches for the computation of discrete distance maps are presented here. The first approach was introduced in [141, 142] and relies on the definition of a mask corresponding to the neighbourhood around each pixel and on which the discrete distance is defined. It will be shown that the implementation of this technique can be achieved using both sequential and parallel architectures (Section 5.2.1). The second approach is based on algorithmic graph theory as presented in Chapter 3. In Section 3.3, it was shown that chamfer distances can be calculated as lengths of shortest paths between vertices in the grid graph. An algorithm will be developed which uses this property and particular properties of shortest path search algorithms for the generation of discrete distance maps (Section 5.2.2). Finally, Section 5.2.3 extends discrete distance mapping to the case of triangular lattices.

5.2.1 Discrete distance masks

The foundation of this technique is the propagation of the discrete distance in P using local distances within the neighbourhood of a pixel. The following proposition states more formally the concept of propagating local distances via neighbouring pixels.

Proposition 5.7:

Given a set of discrete points P and Γ a subset of P, d_{D} is the discrete distance used to compute the distance map of P. Then, for any point $p \in \overset{\circ}{P}$ (i.e. $p \in P \backslash \Gamma$), there exists a point q neighbour of p (i.e. $q \in N_{\mathrm{D}}(p)$) such that $\mathrm{DT_D}(p)$, the discrete distance transformation value at p, is such that $\mathrm{DT_D}(p) = \mathrm{DT_D}(q) + d_{\mathrm{D}}(p,q)$. Moreover, since p and q are neighbours, $l(p,q) = d_{\mathrm{D}}(p,q)$ is the length of the move between p and q. Therefore, for any point $p \notin \Gamma$, q is characterised by $\mathrm{DT_D}(q) = \min\{\mathrm{DT_D}(p) + l(p,q') \ ; \ q' \in N_{\mathrm{D}}(p)\}$.

Proof:
Immediate using the analogy between discrete distances and shortest path lengths presented in Section 3.3.2. □

Both sequential and parallel approaches take advantage of this property to efficiently compute discrete distance maps. A mask is defined which contains local distances within the neighbourhood $p \in P$. This mask is then centred at each point $p \in P$ and local distances are propagated by summing the central value with the corresponding coefficient in the mask.

Definition 5.8: Distance mask

A distance mask of size $(n \times n)$ is an $(n \times n)$ matrix $(m_{k,l})_{k,l}$ where a value $m_{k,l}$ represents the local distance between a pixel $p = (x_p, y_p)$ and a neighbouring pixel $q = (x_p + k, y_p + l)$.

Generally, the mask is centred at pixel p so that its size n is odd and indices k and l are included in $\{-\lfloor \frac{n}{2} \rfloor, \ldots, \lfloor \frac{n}{2} \rfloor\}$.

Figure 5.2 presents such masks for the propagation of local distances. The centre pixel p is shown as a shaded area and represents the centre of the mask ($k = 0$, $l = 0$). The size of the mask is readily given by the type of the neighbourhood considered. Pixels in the neighbourhood of p are associated with the values of the length of the respective moves from p (i.e. move lengths defined in Definition 1.16). The centre pixel is associated with 0. In the case of chamfer distance, these move lengths are denoted a, b, c, *etc.* (see Section 1.4.3). By definition of the neighbourhoods, values represented as a crossed box are not affected in the mask and are considered as infinity (in practice, a large number).

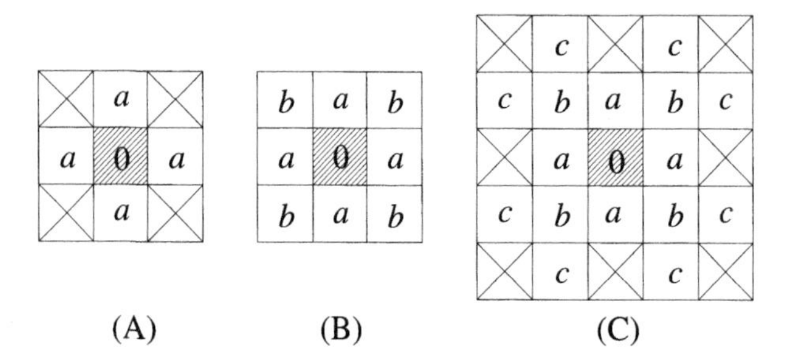

Figure 5.2 Masks for chamfer distance transformations. (A) 4-Neighbourhood. (B) (3×3) mask (8-neighbourhood). (C) (5×5) mask (16-neighbourhood)

In Figure 5.2(A), the mask is based on the 4-neighbourhood and will be used to propagate the d_4 (i.e. City-Block) distance. Similarly, masks presented in Figures 5.2(B) and 5.2(C) will be used to propagate the d_8 ($a = 1$, $b = 1$) or $d_{a,b}$ distances in the 8-neighbourhood and $d_{a,b,c}$ in the 16-neighbourhood, respectively.

The process of calculating discrete distance maps using masks can now be summarised as follows. Given a binary digital image of size $W \times H$, the set Γ is assumed to be known. The discrete distance map is an array of values $(\mathrm{DT_D}(p))_p$ of size $W \times H$ and is calculated by iterative updating of its values until a stable state is reached. The distance map is first initialised as follows (iteration $t = 0$):

$$\mathrm{DT}_\mathrm{D}^{(0)}(p) = \begin{cases} 0 & \text{if } p \in \Gamma \\ +\infty & \text{if } p \notin \Gamma \end{cases} \tag{5.1}$$

Then, at iteration $t > 0$, the mask $(m_{k,l})_{k,l}$ is positioned at a pixel $p = (x_p, y_p)$ and the following updating rule for propagating the distance values from the pixels $q = (x_p + k, y_p + l)$ onto p is used:

$$\mathrm{DT}_\mathrm{D}^{(t)}(p) = \min_{k,l}\{\mathrm{DT}_\mathrm{D}^{(t-1)}(q) + m_{k,l}\ ;\ q = (x_p + k, y_p + l)\} \tag{5.2}$$

The updating process stops when no change is made in the distance map at the current iteration [142]. The sequence in which the mask is positioned on pixels is defined by the type of approach used to implement the distance mapping algorithm. For sequential algorithms, a modified distance mask is swept on the image in two passes. In the case of parallel architecture, at each iteration, the mask is used at every pixel in P to propagate distance outwards from the pixels in Γ.

5.2.1.1 Sequential implementation

The sequential implementation of distance mapping was first introduced in [141] and studied in [12, 13, 166] with the optimisation of move lengths.

For sequential operations, the mask is divided into two symmetric sub-masks. Then, each of these sub-masks is sequentially swept over the initial distance map containing the values $(\mathrm{DT_D}(p))_p^{(0)}$ defined by Equation (5.1) in a forward and backward pass, respectively. This operation is detailed in the following example.

Example: Sequential computation of a discrete distance map

Consider the image shown in Figure 5.3(A) and the (3×3) mask in Figure 5.3(B) (following the results in Section 1.5, $a = 3$ and $b = 4$). The set Γ is defined as the central white pixels and P is the set of all pixels. This mask is first divided in two symmetric sub-masks, as shown in Figures 5.3(C) and (D), where non-set values in the mask are not shown for clarity.

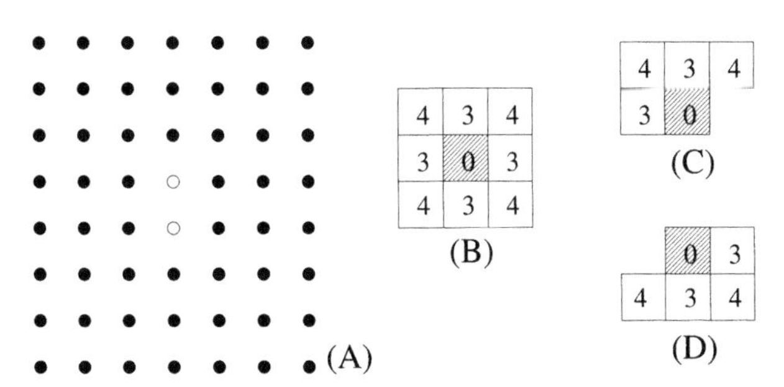

Figure 5.3 Masks for chamfer distance transformations. (A) Original image. (B) (3×3) mask. (C) and (D) Sub-masks for sequential computation

Forward pass: The initial distance map $(\mathrm{DT_D}(p))_p^{(0)}$ is shown in Figure 5.4 (A), where crosses represent infinite values (i.e. corresponding to pixels $p \notin \Gamma$).

The upper sub-mask (Figure 5.3(C)) is positioned at every point of this initial distance map, following the sequence shown in Figure 5.4(B) and the updating rule given by Equation (5.2) is applied. For values corresponding to the border of the image, only coefficients of the mask contained in the distance map are considered. This pass results in the distance map $(\mathrm{DT_D}(p))_p^{(t_1)}$ shown in Figure 5.4(C).

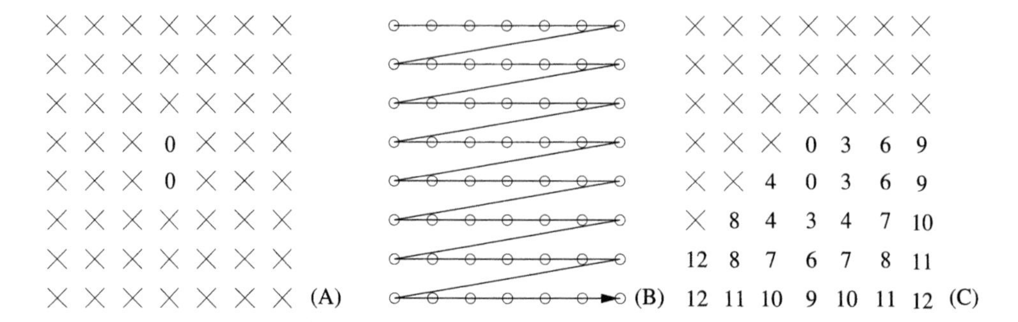

Figure 5.4 Forward pass. (A) Initial distance map. (B) Updating sequence. (C) Temporary discrete distance map $(\mathrm{DT_D}(p))_p^{(t_1)}$ after the forward pass

Backward pass: In a similar fashion, the lower sub-mask (Figure 5.3(D)) is now positioned at every point in the distance map $(\mathrm{DT_D}(p))_p^{(t_1)}$, following the sequence shown in Figure 5.5(A). The updating rule given by Equation (5.2) is applied again.

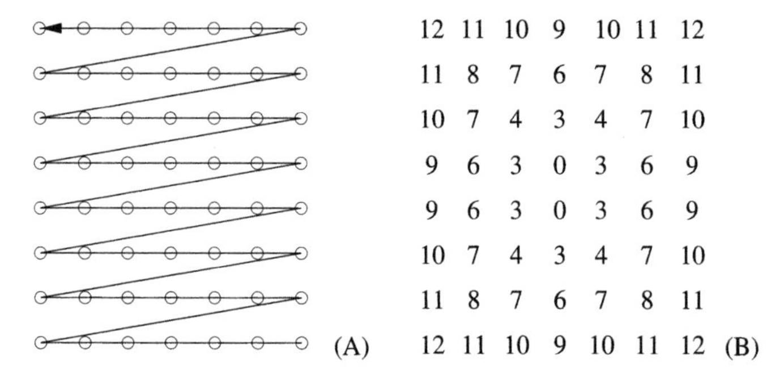

Figure 5.5 Backward pass. (A) Updating sequence. (B) Final discrete distance map

This pass results in the final distance map of the image shown in Figure 5.5(B). One can easily verify that Equation (5.2) is satisfied at every point p in the final distance map. This property derives from Proposition 5.7.

More formally, the sequential algorithm which computes the distance map of a binary digital image of size $W \times H$, using a $(n \times n)$-distance mask $(m_{k,l})_{k,l}$ is presented as Algorithm 5.1.

Clearly, the complexity of this algorithm is $\mathcal{O}(W \times H)$ since the updating rule in Equation (5.2) can be applied in constant time.

5.2.1.2 Parallel implementation

On parallel architectures, each pixel is associated with a processor. The discrete distance map is first initialised using Equation (5.1) and at every iteration, the updating rule given by Equation (5.2) is applied to all pixels. The

[——— Initialisation ———]
for all $p \in \Gamma$ **do**
  $\mathrm{DT}(p) \leftarrow 0$
for all $p \notin \Gamma$ **do**
  $\mathrm{DT}(p) \leftarrow +\infty$
[——— Forward pass ———]
for all pixels $p = (x_p, y_p)$ in the forward sequence, **do**
  $\mathrm{DT}(p) = \min_{k,l}\{\mathrm{DT}(q) + m_{k,l}\ ;\ q = (x_p + k, y_p + l)\}$
[——— Backward pass ———]
for all pixels $p = (x_p, y_p)$ in the backward sequence, **do**
  $\mathrm{DT}(p) = \min_{k,l}\{\mathrm{DT}(q) + m_{k,l}\ ;\ q = (x_p + k, y_p + l)\}$

Algorithm 5.1: Sequential discrete distance mapping algorithm

process stops when no change is made in the discrete distance map by the current iteration. The following example illustrates the parallel computation of a discrete distance map.

Example: Parallel computation of a discrete distance map

Consider the binary digital image presented in Figure 5.3(A). The mask used is the original distance mask shown in Figure 5.3(B) and the initial discrete distance map is shown in Figure 5.4(A) (crosses represent infinite values). Figure 5.6 displays the temporary discrete distance maps obtained during the parallel computation.

(A)

×	×	×	×	×	×	×
×	×	×	×	×	×	×
×	×	4	3	4	×	×
×	×	3	0	3	×	×
×	×	3	0	3	×	×
×	×	4	3	4	×	×
×	×	×	×	×	×	×
×	×	×	×	×	×	×

(B)

×	×	×	×	×	×	×
×	8	7	6	7	8	×
×	7	4	3	4	7	×
×	6	3	0	3	6	×
×	6	3	0	3	6	×
×	7	4	3	4	7	×
×	8	7	6	7	8	×
×	×	×	×	×	×	×

(C)

12	11	10	9	10	11	12
11	8	7	6	7	8	11
10	7	4	3	4	7	10
9	6	3	0	3	6	9
9	6	3	0	3	6	9
10	7	4	3	4	7	10
11	8	7	6	7	8	11
12	11	10	9	10	11	12

Figure 5.6 Temporary discrete distance maps. (A) After first iteration. (B) After second iteration. (C) After third iteration (final discrete distance map)

Clearly, both parallel and sequential process result in the same discrete distance map (Figures 5.5(C) and 5.6(C)).

More formally, the algorithm that computes the discrete distance map of a binary digital image in given in Algorithm 5.2.

At iteration t of the parallel process, the temporary discrete distance map only contains finite values at pixels whose shortest path to a point in Γ is of

```
[——— Initialisation ———]
for all pixels p ∈ P do                              [ One parallel iteration ]
  if p ∈ Γ then
    DT(p) ← 0
  else
    DT(p) ← +∞
[——— Main procedure ———]
repeat
  for all pixels p = (x_p, y_p), do                  [ One parallel iteration ]
    DT(p) = min_{k,l}{DT(q) + m_{k,l} ; q = (x_p + k, y_p + l)}
until no change is made in the distance map.
```

Algorithm 5.2: Parallel discrete distance mapping algorithm

cardinality smaller or equal to t. Moreover, at each iteration, these finite values are correct and will not be updated in a further iteration. Therefore, the complexity of this procedure is given by the maximal value $n_{\max}$ of the cardinality of a shortest path between a point in P and a point in Γ (e.g. in the previous example, $n_{\max} = 3$). Hence, the complexity of the parallel discrete distance mapping procedure is $\mathcal{O}(n_{\max})$.

5.2.2 Graph-theoretic approach

In this section, we summarise the method proposed and detailed in [154], which takes advantage of the optimality of the graph-theoretic shortest path algorithm presented in Chapter 3 to compute efficiently discrete distance maps.

The first step in the computation of discrete distance maps using algorithmic graph theory is to define a procedure which characterises the pixels contained in a discrete disc of a given radius $R \geq 0$ centred at pixel p. This operation will use a label-setting shortest path procedure based on a search list Λ (e.g. see Dial's or Dijkstra's shortest path procedures presented in Section 3.2.1) and the analogy between shortest paths in the grid graph and discrete distances presented in Section 3.3. It was shown in Chapter 3 that such a label-setting algorithm gives a topological ordering to the vertices (i.e. the pixels in the case of a grid graph) traversed during the construction of the shortest path spanning tree. It was also pointed out that it is possible to define a stopping rule based of the distance labels $d(u)$ such that the procedure results in a set of vertices whose shortest path to the vertex s corresponding to the centre pixel p is of length less than or equal to a given value $R \geq 0$. With suitable arc lengths, this property exactly defines $\Delta_{\mathrm{D}}(p, R)$, the discrete disc of radius $R \geq 0$ centred at p for the discrete distance $d_{\mathrm{D}}(.,.)$. More formally, given a set of pixels P, the grid graph $G = (V, A)$ is constructed for the neighbourhood on which the discrete distance

$d_D(.,.)$ is defined. The length $l(u,v)$ of an arc (u,v) between adjacent vertices u and v corresponding to the neighbour pixels q and q', respectively, is chosen as the move length $d_D(q,q')$. A vertex $s \in V$ corresponds to the centre pixel p. The procedure which characterises all vertices $v \in V$ whose shortest path from s is of length smaller than $R \geq 0$ can be summarised in Algorithm 5.3.

```
[——— Initialisation ———]
for all vertices q ∈ V do
    d(q) ← +∞
d(s) ← 0
Initialise the search list Λ
u ← s                          [ The search starts from the centre vertex s ]
[——— Main procedure ———]
while (d(u) ≤ R) and (the search list Λ is non-empty) do
    Set the label d(u) as permanent
    for all vertices v in the forward star of u do
        d(v) ← min{d(v), d(u) + l(u,v)}    [ Updates the temporary label of v ]
        Store v in the search list Λ
    Get the next vertex u from the search list Λ
[——— Discrete disc characterisation ———]
Δ_D(p,R) is the set of pixels corresponding to vertices u whose label d(u) is
permanent.
```

Algorithm 5.3: Graph-theoretic algorithm for the characterisation of discrete discs

For an extended presentation of label-setting shortest path algorithms (e.g. for details about the search list Λ used in such algorithms), the reader is referred to Section 3.2.1. Note that, if P does not totally contain $\Delta_D(p,R)$, the search list Λ may be emptied before all temporary labels have become greater than R. Such a case is considered in Algorithm 5.3 by adding an extra exiting condition in the main "**while**" loop.

Clearly, following the study of shortest path search algorithms in Section 3.2.1, the choice of whether a unique list Λ (e.g. Dijkstra's algorithm) or a set of buckets Λ_i (e.g. Dial's algorithm) is to be used is determined by the value of arc lengths. In the case of chamfer distances, the values of local distances (i.e. move lengths) are typically small (e.g. $a = 3$ and $b = 4$). In this case, an implementation using buckets following Dial's shortest path algorithm (Algorithm 3.4) will be computationally efficient. G is a sparse graph. Therefore, if buckets (i.e. different search lists Λ_i) are used, the complexity of this procedure is typically $\mathcal{O}(R^2)$, since all pixels that need to be investigated are contained in a square of size $R \times R$. In other words, the complexity of this procedure used on a sparse grid graph and when using different search lists (i.e. buckets) is at worst of order $\mathcal{O}(N)$, where $N = |V|$ is the number of vertices (i.e. the number of pixels in P).

The graph-theoretic algorithm which computes the discrete distance map of a set of pixel P makes use of this procedure [154]. The outline of this algorithm is as follows. Given a set P, a subset Γ and a discrete distance function $d_D(.,.)$, the grid graph $G = (V, A)$ of P is supposed to be constructed. Moreover, the set of vertices corresponding to pixels in Γ is denoted V_Γ (i.e. $V_\Gamma \subseteq V$). Values in the discrete distance map are first initialised via Equation (5.1). Then, discrete discs are expanded from each vertex $s \in V_\Gamma$ using the main "**while**" loop in Algorithm 5.3. At each vertex $u \in V$, a number of such discrete discs overlap. This is sketched in Figure 5.7 where shaded circles represent the border of discrete discs expanded from the border vertices $s \in V_\Gamma$.

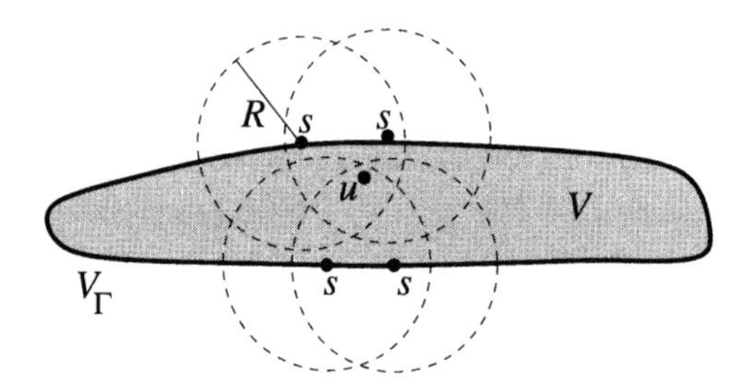

Figure 5.7 Sketch for the graph-theoretic distance mapping

Since the minimal distance label is considered at each vertex u, at the end of the algorithm, each vertex u receives a distance label $d(u)$ which is the length of the shortest path from u to its closest vertex in V_Γ. Clearly, such a label $d(u)$ defines $DT_D(p)$, the value in the discrete distance map at pixel p corresponding to vertex u. Algorithm 5.4 summarises such a procedure.

Remark 5.9:

The value $R \geq 0$ of the radii of the discrete discs expanded in V (i.e. in P) need not to be known a priori and can be estimated as one half of the diameter of the grid graph (e.g. see Chapter 6).

The complexity of this procedure depends on two factors. Firstly, $|V_\Gamma|$ discrete discs are to be expanded, where $|V_\Gamma|$ is the number of border vertices (i.e. the number of pixels in Γ). Therefore, the complexity of this procedure depends on this number, which in turn depends on the set Γ considered (note that, in general, $|\Gamma| \ll |P|$). Secondly, if the common radius R for the discrete discs is chosen accurately, the complexity of the procedure that expands discrete discs falls below $\mathcal{O}(N)$, assuming the use of buckets Λ_i. Therefore, the global complexity of Algorithm 5.4 can be expressed as $\mathcal{O}(|\Gamma|.R^2)$. Note that, if R is chosen as a randomly large value, all $N = |V|$ vertices need to be investigated for each border vertex s.

```
[——— Initialisation ———]
for all vertices q ∈ V \ V_Γ do
   d(q) ← +∞
for all vertices q ∈ V_Γ do
   d(s) ← 0
[——— Main procedure ———]
for all vertices s ∈ V_Γ do
   Initialise the search list Λ
   u ← s                    [ A discrete disc is expanded from each s ∈ V_Γ ]
   while ((du) ≤ R) and (the search list Λ is non-empty) do
      Set the label d(u) as permanent
      for all vertices v in the forward star of u do
         d(v) ← min{d(v), d(u) + l(u,v)}  [ Updates the temporary label of v ]
         Store v in the search list Λ
      Get the next vertex u from the search list Λ
[——— Discrete distance map characterisation ———]
for all permanently labelled vertices u in V do
   DT(p) ← d(u), where vertex u corresponds to pixel p in G.
```

Algorithm 5.4: Graph-theoretic algorithm for the discrete distance mapping

Remark 5.10:

In the case of binary digital line images, the value of R is typically small and this algorithm proves efficient for this class of images (see Section 7.3).

The graph-theoretic approach presented here offers different advantages which we summarise in the proposition below.

Proposition 5.11:

Using algorithmic graph theory for the computation of discrete distance maps allows for the following.

(i) *Using the concept of the grid graph, connected components in an image are readily identified (see Chapter 6). Therefore, this approach allows for computing the distance maps of different connected components separately. By contrast, the approach based on distance masks and presented in Section 5.2.1 does not permit such a processing approach.*

(ii) *Similarly, this approach readily allows for the characterisation of the discrete Voronoi diagram of the set P with respect to points in the set Γ using* *the definition of a pixel* ref(p) *for every pixel in* P *(see also Proposition 5.4(ii)).*

(iii) *Finally, this approach allows for the computation of constrained discrete distance maps (see Section 5.4.1). It can be shown that the constrained distance mapping problem is a trivial extension of the distance mapping problem when algorithmic graph theory is used.*

These properties will be used for the development of further algorithms in the following chapters and will be detailed in Section 5.4.

5.2.3 Triangular lattices

Discrete distance transformations on triangular lattices (i.e. on sets of pixels resulting from a hexagonal partitioning) are detailed in [12, 14]. The aim is to obtain a discrete distance map using discrete distances defined on triangular lattices in Section 1.2.1 (e.g. d_6 distance). This technique is typically the same as that presented in Section 5.2.1, where the mask is adapted to the triangular neighbourhood defining the d_6 distance (Figure 5.8(A)). This mask is divided into two parts as shown in Figure 5.8(B) and swept over the image in two passes.

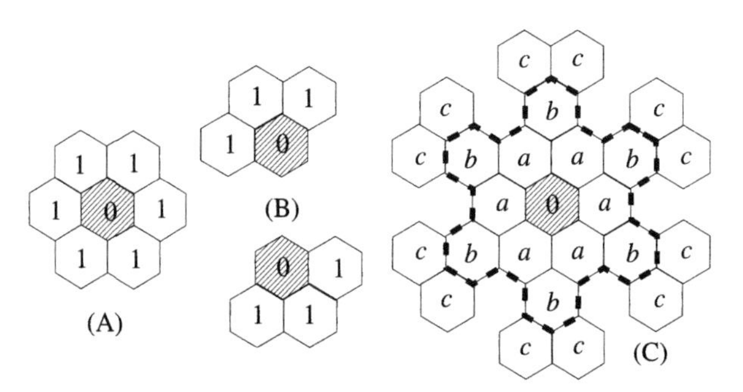

Figure 5.8 Distance mapping on triangular lattices. (A) Mask in the 6-neighbourhood. (B) Sub-masks. (C) Extended mask

Extended masks are also proposed here (e.g. see Figure 5.8(C)) and optimum coefficients are proposed. The first extension of the 6-neighbourhood leads to the definition of the move lengths a and b, by analogy with chamfer distances. This mask is delimited by a bold dashed line in Figure 5.8(C). Optimal integer values $a = 3$ and $b = 5$ are proposed for the generation of discrete distance maps on triangular lattices. Similarly, a second extension of this neighbourhood leads to the definition of an extra move whose length is denoted c and optimal values for these coefficients are given as $a = 8$, $b = 14$ and $c = 21$.

We emphasise the fact that the graph-theoretic approach can readily be used for the computation of discrete distance maps on triangular lattices. Via the definition of the grid graph that corresponds to a given neighbourhood on triangular lattice and suitable move lengths (e.g. of lengths a, b and c defined above), the algorithms presented in Section 5.2.2 can be applied. Since Algorithms 5.3 and 5.4 depend only on the grid graph used, no change is necessary for these algorithms to be used as discrete distance mapping algorithms on triangular lattices.

5.3 Euclidean distance transformations

The problem of Euclidean distance mapping is presented in this section as a solution of typical inconsistencies that arise when using discrete distance maps. These inconsistencies are first briefly reviewed and a formulation of the Euclidean distance mapping is presented. This problem is shown to be non-trivial for practical applications. In this respect, approximate and exact solutions for the Euclidean distance mapping problem are both reviewed using the same type of approaches as in the previous section.

Discrete distance mapping typically results in a set of small integer values which represent the discrete distance for a point $p \in \Gamma$ to its closest point in $\Gamma \subseteq P$. This processing is efficient in terms of both computation and storage since only integer arithmetic is involved and small integer values can easily be represented on a computer (e.g. see Section 5.4). However, following the study on compatibility between discrete and Euclidean distances presented in Section 1.5, discrete distance maps are not accurate enough in some cases. Typically, the non-invariance of discrete distance against geometric operations such as scaling and rotation may dramatically affect discrete distance maps. Moreover, it has been shown in [42,92] that correct Euclidean distance maps can be deduced from discrete distance maps for limited distance values only (see also Section 1.5). Beyond a certain limit, Euclidean distance values deduced from discrete ones result in erroneous values.

The solution to these inconsistencies is to use Euclidean distances directly for the distance mapping. Therefore, Euclidean distance mapping refers to the problem of operating a distance transformation defined by Definition 5.1, where the distance function $d(.,.)$ is the Euclidean distance function $d_{\mathrm{E}}(.,.)$ (see Definition 1.25). However, three main shortcomings arise when using such a continuous distance.

Firstly, Euclidean distance is a global function on the set of pixels P. Unlike discrete distances it is not implicitly defined by local distances within a neighbourhood. The propagation of Euclidean distance therefore becomes a non-trivial problem. Secondly, Euclidean distance calculations involve square root operations and, more generally, non-integer arithmetic and may therefore not be efficiently computed. Moreover, as pointed out in [142], neither $d_{\mathrm{E}}^2(.,.)$, nor an integer mapping of the Euclidean distance (e.g. $\lfloor d_{\mathrm{E}}(.,.) \rfloor$, $\lceil d_{\mathrm{E}}(.,.) \rceil$ and

$[d_E(.,.)]$) satisfy the metric conditions in Definition 1.15. Therefore, alternative solutions are to be defined for overcoming the complexity of calculations involved in Euclidean distance mapping. Finally, the non-integer nature of Euclidean distance maps make them unsuitable for computer storage. Approximating final Euclidean values with integer ones via truncating or squaring operations may also involve large integer values which are themselves unsuitable for storage.

The following approach is generally used for solving both these problems. Instead of storing a unique value at each pixel p that represents the Euclidean distance between p and its closest point s in Γ, a pair of integer values $\mathrm{EDT}(p) = (\delta_x(p), \delta_y(p))$ is stored in the Euclidean distance map. Integer values $\delta_x(p)$ and $\delta_y(p)$ represent the displacements from p to its closest point $s \in \Gamma$ on the x- and y-axis, respectively. Euclidean distance maps are said to be signed or unsigned depending on whether values $\delta_x(p)$ and $\delta_y(p)$ are calculated as signed or unsigned integers. We mostly detail the case of unsigned Euclidean distance map since absolute displacement values facilitate storage and still allow for the computation of the final Euclidean distance map values. For further details on signed Euclidean distance maps, the interested reader is referred to [123,176]. Note that, in most references on Euclidean distance mapping, the generation of both signed and unsigned maps is considered. Here, it is assumed that $\delta_x(p) = |x_s - x_p|$ and $\delta_y(p) = |y_s - y_p|$, where s is the closest point to p in Γ. Then, $\mathrm{DT_E}(p)$, the Euclidean distance transformation value at pixel p, can be retrieved using the definition of Euclidean distance (Definition 1.25),

$$\mathrm{DT_E}(p) = \|\mathrm{EDT}(p)\|_2 = \sqrt{\delta_x(p)^2 + \delta_y(p)^2}$$

This formulation allows for the propagation of the Euclidean distance and therefore sequential or parallel implementations similar to that detailed in the case of discrete distances can be achieved [28, 90, 123, 173]. However, in some cases, only approximations of the Euclidean distance map are obtained (e.g. see [28]), while other approaches allow for the computation of exact Euclidean distance maps (e.g. see [90, 123, 173]).

The principle of Euclidean distance mapping can now be summarised as follows. The Euclidean distance map is first initialised in a similar way as in Equation (5.1), using Equation (5.3).

$$\mathrm{EDT}^{(0)}(p) = \begin{cases} (0,0) & \text{if } p \in \Gamma \\ (+\infty, +\infty) & \text{if } p \notin \Gamma \end{cases} \tag{5.3}$$

Then, at iteration t, a value $\mathrm{EDT}^{(t)}(p)$ in the Euclidean distance map is updated in a sequence which depends on the implementation (e.g. sequential or parallel), using the updating rule given by Equation (5.4).

$$\mathrm{EDT}^{(t)}(p) = \mathrm{EDT}^{(t-1)}(q) + (|x_q - x_p|, |y_q - y_p|) \quad (5.4)$$

where q characterises the minimum value of $\|\mathrm{EDT}^{(t-1)}(q')+(|x_{q'}-x_p|, |y_{q'}-y_p|)\|_2$ for all q' in a given neighbourhood of p.

Throughout this section, different approaches are presented that allow for the computation of approximate and exact Euclidean distance maps. Both advantages and drawbacks of these techniques are detailed. Section 5.3.1 presents the propagation of Euclidean distances using the concept of distance masks introduced in the case of discrete distances. Both sequential and parallel implementations are presented in this section. Continuing with the context of algorithmic graph theory, Section 5.3.2 presents an algorithm for solving exactly the Euclidean distance mapping problem. Extra information obtained using this approach is highlighted. Finally, Section 5.3.3 summarises other approaches taken for the computation of Euclidean distance maps.

5.3.1 Euclidean distance masks

This approach was first proposed in [28] as a pioneering solution to the Euclidean distance mapping problem. It is shown that, except in some particular configuration of points in Γ, one can obtain correct Euclidean distance maps by using this approach. We introduce the computation of Euclidean distance maps using the concept of distance masks with an outline of this method and by presenting a special case in which inconsistencies arise.

As mentioned earlier, Euclidean distance maps are found via the storage of a pair of integer values $\mathrm{EDT}(p) = (\delta_{\mathrm{x}}(p), \delta_{\mathrm{y}}(p))$ at each pixel p which represents the displacement from p to its closest point in Γ. Euclidean distance masks follow this approach. Instead of containing local distances within a neighbourhood, a Euclidean distance mask will contain local displacements made within this neighbourhood. An example of such a mask based on the 8-neighbourhood is shown in Figure 5.9(A). This mask is centred on the pixel as indicated by the shaded centre box containing the displacements $(0, 0)$.

5.3.1.1 A basic sequential algorithm

The sequential algorithm that realises the computation of the Euclidean distance map is then similar to that presented in the case of discrete distances. Values $\mathrm{EDT}^{(0)}(p)$ in the discrete distance map are first initialised using Equation (5.3) as shown in Figure 5.9(B), where crosses represent infinite values (i.e. $(+\infty, +\infty)$) and brackets are dropped for clarity. Then, in two passes (one upward and one downward pass), the values $\mathrm{EDT}^{(t)}(p)$ are successively updated using the Euclidean distance mask and Equation (5.4).

1,1	0,1	1,1
1,0	0,0	1,0
1,1	0,1	1,1

(A)

×	×	×	×	×	×	×
×	×	×	×	×	×	×
×	×	×	×	×	×	×
×	×	×	0,0	×	×	×
×	×	×	0,0	×	×	×
×	×	×	×	×	×	×
×	×	×	×	×	×	×
×	×	×	×	×	×	×

(B)

Figure 5.9 (A) Euclidean distance mask in the 8-neighbourhood space. (B) Initial Euclidean distance map $(\text{EDT}(p))_p^{(0)}$

Figures 5.10 and 5.11 illustrate such a sequential Euclidean distance mapping process applied on the binary image presented in Figure 5.3(A). At each pixel p a pair of integer values indicates the value of $\text{EDT}(p)$.

Upward sequential pass: The distance map is scanned line by line, from the bottom line to the top line using three different sub-masks. For each line, all three sub-masks whose union is the lower half of the complete Euclidean mask shown in Figure 5.9(A) are used successively (see Figures 5.10(A) to (C)). At each step, the updating rule given in Equation (5.4) is applied for the neighbourhood of p defined by the sub-mask considered. The sub-mask shown in Figure 5.10(A) is first applied successively from left to right at each pixel of the current line. In a similar way, the sub-mask shown in Figure 5.10(B) is applied successively from left to right at each pixel of the current line. Finally, the sub-mask shown in Figure 5.10(C) is applied successively from right to left at each pixel of the current line. This pass results in a temporary Euclidean distance map containing the values $\text{EDT}^{(t_1)}(p)$ shown in Figure 5.10(D).

Clearly, the upward pass propagates correct Euclidean distance values from the lowest line that contains a point in Γ. In other words, finite values $\text{EDT}^{(t_1)}(p)$ in the temporary distance map are correct and are therefore not to be updated in a further iteration.

Downward sequential pass: The Euclidean distance map is again scanned line by line from top to bottom. For each line, the upper half of the mask shown in Figure 5.9(A) is used via a decomposition in three sub-masks (Figures 5.11(A) to (C). The first sub-mask (Figure 5.11(A)) is applied on each pixel of the current line from left to right. Similarly, the sub-mask shown in Figure 5.11(B) is applied on each pixel of the current line from left to right. Finally, the sub-mask shown in Figure 5.11(C) is applied on each pixel of the current line from right to left. This pass results in the final Euclidean distance mask shown in Figure 5.11(D).

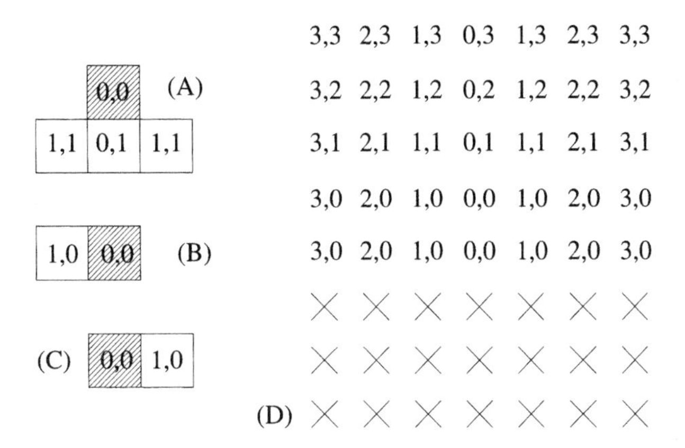

Figure 5.10 Upward pass for the sequential algorithm. (A) to (C) Euclidean distance sub-mask. (D) Temporary Euclidean distance map $(\mathrm{EDT}(p))_p^{(t_1)}$

One can easily verify that Equation (5.4) holds at each point of this distance map.

Clearly, the complexity of this procedure applied on a binary image of size $W \times H$ is $\mathcal{O}(W \times H)$ since the updating rule can be applied in constant time.

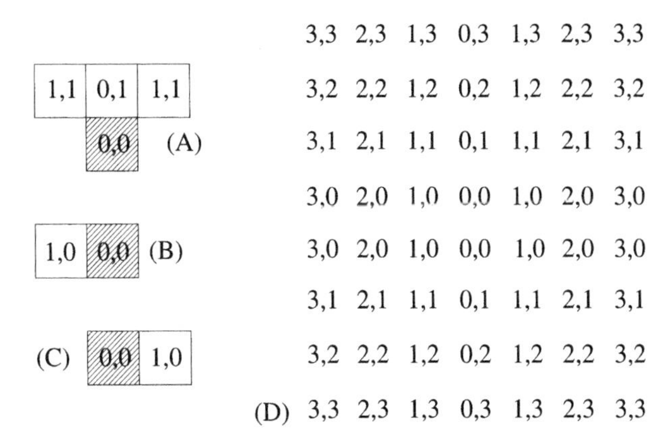

Figure 5.11 Downward pass for the sequential algorithm. (A) to (C) Euclidean distance sub-mask. (D) Final Euclidean distance map $(\mathrm{EDT}(p))_p$

Inconsistencies: Typical problems that arise during the generation of Euclidean distance maps using this mask can be characterised by the fact that some configurations of points in Γ may prevent the propagation of correct Euclidean distances to certain points. This is illustrated by the following example, where the 8-neighbourhood is considered.

Example: Incorrectness of a Euclidean distance map

Consider the part of a binary image shown in Figure 5.12. The set Γ is composed of the three white pixels s, s_0 and s_1. Consider the case of pixel p highlighted in the image. Pixel p belongs to the Voronoi cell associated with the pixel $s \in \Gamma$ and the value of the Euclidean distance map at p should be of the form $(|x_s - x_p|, |y_s - y_p|)$. The dashed lines represent the border of the Voronoi cell associated with s.

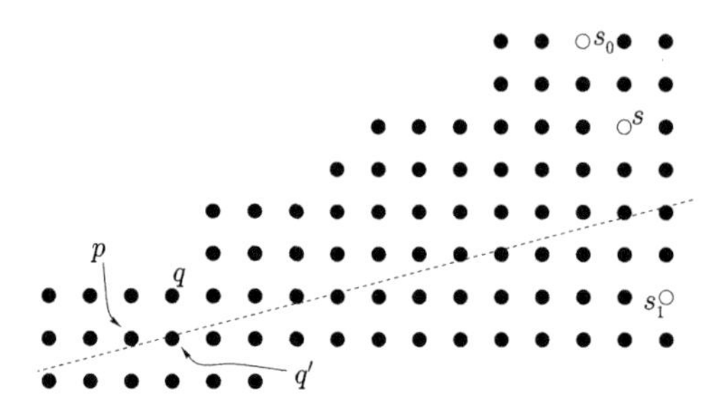

Figure 5.12 A case where the Euclidean distance cannot be propagated onto p

Now, p is not 8-connected with any pixel in the Voronoi cell of s (i.e. q and q' do not belong to this cell). Such a configuration clearly prevents a correct propagation of the Euclidean distance from s to p. The value of the Euclidean distance map at p will be dictated by the Euclidean distance from either s_0 or s_1. In other words, the propagation of the Euclidean distance will be done through q or q' whose Euclidean distance transformation is defined according to its distance to s_0 and s_1, respectively

This type of problem can be summarised by the fact that, during the Euclidean distance propagation, a "barrier" is formed which prevents the propagation of correct distance values to some other pixels. This is the case in the previous example, where q and q' are associated with Euclidean distance values that prevent the propagation of a correct Euclidean distance from s to p.

However, it should be noted that this type of configuration arises in rare case only and, although incorrect, the value associated with the pixel in question (e.g. p in the previous example) can be considered as a close approximation of the correct Euclidean distance transformation value at this pixel. In this respect, this approach may be satisfactory for some applications since it is simple and can be implemented efficiently.

5.3.1.2 Parallel and contour processing approaches

Two approaches are classically taken for solving inconsistencies created by the propagation of Euclidean distances in the distance map.

Firstly Yamada [173] proposed a parallel equivalent of the previous algorithm, where the Euclidean distance values are updated in parallel at every pixel at each iteration. Equations (5.3) and (5.4) are used for the initialisation and updating steps, respectively. It is shown that the parallel implementation of this algorithm prevents the creation of "barriers" mentioned earlier since all pixels are updated simultaneously. Therefore, this algorithm results in an error-free Euclidean distance map of the set P in question.

However, where parallel implementation is not possible, it is of interest to define a sequential algorithm that results in error-free Euclidean distance maps. Such an algorithm is proposed in [123,124,127,128], based on a technique referred to as *contour processing neighbourhood operation* first introduced in [164].

The idea behind the contour processing technique is twofold. Firstly, in operations such as the propagation of distance over a set of pixels, at each stage, only a restricted set of pixels would truly benefit from the application of the updating rule. It has been noted that, in both sequential and parallel distance mapping algorithms, the value at a pixel is mostly updated during one iteration (i.e. the iteration at which the propagation front reaches that pixel). Therefore, some computational effort is made redundant by such a characterisation. By maintaining a list of pixels (called the *contour set*) on which updating operations will be performed, this redundancy can be removed. Clearly, the contour set can be characterised by the propagation front of the Euclidean distance in the distance map. Moreover, it has been noted in [123] that such a list allows for the emulation of a parallel process using sequential operations. Applying this principle to distance map generation will therefore guarantee the correctness of the resulting Euclidean distance map and may also improve the computational efficiency of the basic algorithm proposed earlier.

The algorithm based on contour processing operation is summarised in Algorithm 5.5. Two search lists Λ_1 and Λ_2 containing elements of the form $[p, (\delta_x, \delta_y)]$ are used. At each stage of the main procedure, the list Λ_1 contains pixels p and values (δ_x, δ_y) which may define a smaller value for $\mathrm{DT_E}(p)$. If this is the case, $\mathrm{EDT}(p) = (\delta_x(p), \delta_y(p))$ is updated to the values (δ_x, δ_y). Moreover, each time the Euclidean distance label $\mathrm{EDT}(p)$ of a pixel p is updated, neighbours q of p contained in a portion of the plane defined by continuity by the new value of $\mathrm{EDT}(p)$ are stored in Λ_2 to be tested at the next propagation stage (i.e. **if**s in Algorithm 5.5). This scheme is illustrated in Figure 5.13, where the direction of propagation for the next step depends on the position of p (shaded discs), relatively to s (black disc).

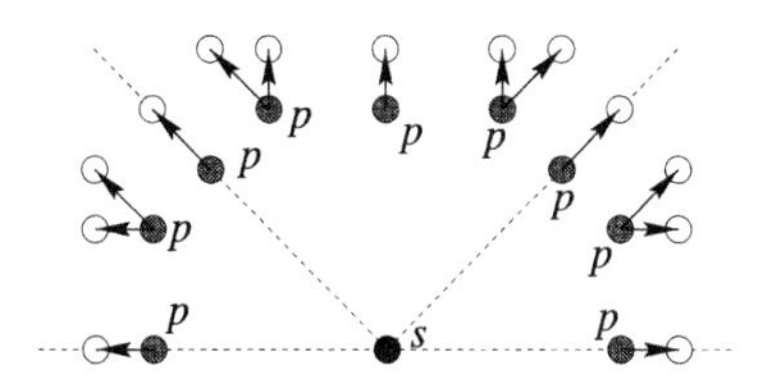

Figure 5.13 Propagation of the Euclidean distance using contour processing

For each case, the direction of propagation is represented by arrows (other cases can be found by symmetry). The interested reader will compare this restriction with the definition of "influence cones" introduced in [162]. Therefore, the lists Λ_1 and Λ_2 alternatively contain the propagation front for the Euclidean distance and create a buffer which allows for the simultaneous updating of the Euclidean distance labels. The algorithm thus guarantees the correctness of the resulting Euclidean distance map.

Subroutine TEST(q,(δ_x, δ_y))
$q' \leftarrow q + (\delta_x, \delta_y)$ [i.e. q' is such that $x_{q'} - x_q = \delta_x$ and $y_{q'} - y_q = \delta_y$]
if $\delta_x(q')^2 + \delta_y(q')^2 > (\delta_x(q) + \delta_x)^2 + (\delta_y(q) + \delta_y)^2$ **then**
  Store $[q', (\delta_x(q) + \delta_x, \delta_y(q) + \delta_y)]$ in Λ_2
end of subroutine TEST

[——— Initialisation ———]
for all pixels s in Γ **do**
  for all pixels p, 8-neighbours of s **do**
    TEST(s,$(x_p - x_s, y_p - y_s)$)
Switch Λ_1 and Λ_2
[——— Main procedure ———]
while list Λ is non-empty **do**
  for all $[p, (\delta_x, \delta_y)]$ in Λ_1 **do**
    if $\delta_x(p)^2 + \delta_y(p)^2 > \delta_x^2 + \delta_y^2$ **then**
      EDT(p) $\leftarrow (\delta_x, \delta_y)$
      if $\delta_x(p) = 0$, **then** TEST(p,$(0, \text{sign}(\delta_y(p)))$)
      else if $\delta_y(p) = 0$, **then** TEST(p,$(\text{sign}(\delta_x(p)), 0)$)
      else if $|\delta_x(p)| = |\delta_y(p)|$, **then** TEST($p$,$(\text{sign}(\delta_x(p)), \text{sign}(\delta_y(p)))$)
      else if $|\delta_x(p)| > |\delta_y(p)|$, **then**
        TEST(p,$(\text{sign}(\delta_x(p)), \text{sign}(\delta_y(p)))$); TEST($p$,$(\text{sign}(\delta_x(p)), 0)$)
      else TEST(p,$(\text{sign}(\delta_x(p)), \text{sign}(\delta_y(p)))$); TEST($p$,$(0, \text{sign}(\delta_y(p)))$)
  Switch Λ_1 and Λ_2
  Empty Λ_2
[——— Euclidean distance map characterisation ———]
The Euclidean distance map is given by the set of values EDT(p)

Algorithm 5.5: Euclidean distance mapping based on contour processing

The complexity of this algorithm applied to an image of size $W \times H$ is given as $\mathcal{O}(W \times H)$, since this algorithm updates the Euclidean distance label of each pixel only once. Therefore, this algorithm is optimal [123].

5.3.2 Graph-theoretic approach

This approach continues with the context of algorithmic graph theory already used in Section 5.2.2 for the generation of discrete distance maps. The algorithm presented here takes advantage of special properties of the Euclidean distance to obtain an exact solution to the Euclidean distance mapping problem [90]. It is to be compared with the contour processing distance transformations proposed in the above section where the principle of propagating local distance on a restricted area only is maintained. However, the use of algorithmic graph theory shows more formally the robustness of this type of approach and allows for operating an optimal distance propagation (see Section 3.2.1).

Given a set of pixels P and a subset Γ, the 8-grid graph $G = (V, A)$ of P is constructed and the set of vertices in V corresponding to pixels in Γ is denoted V_Γ. A similar approach to that presented in Section 5.2.2 will be used. More precisely, discs are expanded from each vertex s in V_Γ and the minimal distance value at each point is stored as the distance transformation at each point. However, two major differences can readily be stated between the discrete and Euclidean distance mapping algorithms. Firstly, arcs lengths are to be calculated in a different way, in order to include the continuous property of the Euclidean distance. Secondly, the updating rule is to be applied in a different way for overcoming configurations of points such as those presented in Figure 5.12.

The principles behind the calculation of discrete and continuous distances using shortest path lengths are detailed in Section 3.3.2. For the sake of simplicity, vertices are identified with their corresponding points and arcs are identified with moves they correspond to. Consider two vertices $s \in V$ and $u \in V$ such that the shortest path between s and u is of length $d_{a,b}(s,u) = a.k_{\mathrm{a}}(s,u) + b.k_{\mathrm{b}}(s,u)$, where $k_{\mathrm{a}}(s,u)$ and $k_{\mathrm{b}}(s,u)$ represent the number of a- and b-moves on this path, respectively (see Section 3.3.2). At this stage, the grid graph is considered to be complete so that $d_{\mathrm{E}}(s,u) = \sqrt{(k_{\mathrm{a}}(s,u) + k_{\mathrm{b}}(s,u))^2 + k_{\mathrm{b}}(s,u)^2}$. A vertex v, neighbour of u can be reached from u by either an a-move or a b-move. Now, if u is on the shortest path from s to v, clearly, $d_{a,b}(s,v) = d_{a,b}(s,u) + d_{a,b}(u,v)$ (see Proposition 3.28 for more details). Two cases are then distinguished:

- If (u,v) is an a-move, $d_{a,b}(u,v) = a$ and $d_{a,b}(s,v) = a.k_{\mathrm{a}}(s,v) + b.k_{\mathrm{b}}(s,v) = a(k_{\mathrm{a}}(s,u)+1) + b.k_{\mathrm{b}}(s,u)$. In this case,

$$d_{\mathrm{E}}(s,v) = \sqrt{((k_{\mathrm{a}}(s,u)+1) + k_{\mathrm{b}}(s,u))^2 + k_{\mathrm{b}}(s,u)^2}$$

and the change in the square of the Euclidean distance between u and v can be calculated as

$$d_{\mathrm{E}}(s,v)^2 - d_{\mathrm{E}}(s,u)^2 = 2(k_{\mathrm{a}}(s,u) + k_{\mathrm{b}}(s,u)) + 1 \qquad (5.5)$$

- If (u, v) is a b-move, $d_{a,b}(u, v) = b$ and $d_{a,b}(s, v) = a.k_{\mathrm{a}}(s, v) + b.k_{\mathrm{b}}(s, v) = a.k_{\mathrm{a}}(s, u) + b(k_{\mathrm{b}}(s, u) + 1)$. In this case,

$$d_{\mathrm{E}}(s, v) = \sqrt{(k_{\mathrm{a}}(s, u) + (k_{\mathrm{b}}(s, u) + 1))^2 + (k_{\mathrm{b}}(s, u) + 1)^2}$$

and the change in the square of the Euclidean distance between u and v can be calculated as

$$d_{\mathrm{E}}(s, v)^2 - d_{\mathrm{E}}(s, u)^2 = 2k_{\mathrm{a}}(s, u) + 4k_{\mathrm{b}}(s, u) + 2 \tag{5.6}$$

The change in the square of the Euclidean distance between two neighbour vertices will form the basis for the propagation of Euclidean distances in the grid graph [90, 92]. Given a vertex s as origin, the Euclidean distance will be propagated correctly along an 8-path from a vertex s if and only if the following recursive formulae are satisfied for any neighbouring vertices u and v on this path

- $d_{\mathrm{E}}(s, s)^2 = 0$, $k_{\mathrm{a}}(s, s) = 0$, $k_{\mathrm{b}}(s, s) = 0$.
- If (u, v) is an a-move: $d_{\mathrm{E}}(s, v)^2 = d_{\mathrm{E}}(s, u)^2 + 2(k_{\mathrm{a}}(s, u) + k_{\mathrm{b}}(s, u)) + 1$, $k_{\mathrm{a}}(s, v) = k_{\mathrm{a}}(s, u) + 1$, $k_{\mathrm{b}}(s, v) = k_{\mathrm{b}}(s, u)$.
- If (u, v) is a b-move: $d_{\mathrm{E}}(s, v)^2 = d_{\mathrm{E}}(s, u)^2 + 2k_{\mathrm{a}}(s, u) + 4k_{\mathrm{b}}(s, u) + 2$, $k_{\mathrm{a}}(s, v) = k_{\mathrm{a}}(s, u)$, $k_{\mathrm{b}}(s, v) = k_{\mathrm{b}}(s, u) + 1$.

Therefore, given a vertex s as origin and two neighbouring vertices u and v, adaptive move lengths are defined as follows.

- If (u, v) is an a-move: $l(u, v) = d_{\mathrm{E}}(s, v)^2 - d_{\mathrm{E}}(s, u)^2 = 2(k_{\mathrm{a}}(s, u) + k_{\mathrm{b}}(s, u)) + 1$.
- If (u, v) is a b-move: $l(u, v) = d_{\mathrm{E}}(s, v)^2 - d_{\mathrm{E}}(s, u)^2 = 2k_{\mathrm{a}}(s, u) + 4k_{\mathrm{b}}(s, u) + 2$.

Using these move lengths in an algorithm similar to that presented as Algorithm 5.3, one can readily expand Euclidean discs from any vertex on the grid graph. The algorithm which expands a Euclidean disc of radius $R \geq 0$ and centred at vertex s is presented as Algorithm 5.6 and is again based on a label-setting shortest path procedure. It is important to note that, since the arc lengths are to be calculated while operating the shortest path search, the maximal value of these arc lengths is not bounded *a priori*. Therefore, Dial's shortest path procedure may not be efficient for such a search. The number of buckets to be used in this procedure is bounded only by the radius of the disc to expand. It should therefore be preset to a large value, thus increasing the complexity of the procedure and degrading the performance. In this case, the label-setting shortest

```
[———— Initialisation ————]
for all vertices q ∈ V do
   d(q) ← +∞
d(s) ← 0; k_a(s,s) ← 0; k_b(s,s) ← 0
Initialise the search list Λ
u ← s                          [ The search starts from the centre vertex s ]
[———— Main procedure ————]
while (d(u)^2 ≤ R^2) and (the search list Λ is non-empty) do
   Set the label d(u) as permanent
   for all vertices v in the forward star of u do
      if (u,v) is an a-move then
         l(u,v) ← 2(k_a(s,u) + k_b(s,u)) + 1
         δk_a ← 1; δk_b ← 0
      else                                         [ (u,v) is a b-move ]
         l(u,v) ← 2k_a(s,u) + 4k_b(s,u) + 2
         δk_a ← 0; δk_b ← 1
      if d(v) > d(u) + l(u,v) then      [ Updates the temporary label of v ]
         d(v) ← d(u) + l(u,v)
         k_a(s,v) ← k_a(s,u) + δk_a; k_b(s,v) ← k_b(s,u) + δk_b
         Store v in the search list Λ
   Get the next vertex u from the search list Λ
[———— Discrete disc characterisation ————]
Δ_E(s,R) is the set of vertices u whose label d(u) is permanent.
```

Algorithm 5.6: Euclidean discs on the 8-grid graph

path procedure presented as Dijkstra's shortest path algorithm (Algorithm 3.3 in Section 3.2.1) is proved to be more efficient to operate the shortest path search.

The validity of Algorithm 5.6 relies on the fact that arcs lengths are defined such that, at each stage, Equations (5.5) and (5.6) are valid. Moreover, since $d_E(s,v) > d_E(s,u)$ implies that $d_E(s,v)^2 > d_E(s,u)^2$, the distance label $d(u)$ represents at each stage the square of the Euclidean distance between u and the origin vertex s.

Since G is a sparse graph with positive arc lengths and since the search is operated within a square of size $R \times R$, the complexity of this procedure can be expressed as $\mathcal{O}(R^2 \log R^2)$ (see Section 3.2.1 for details on complexities of shortest path search algorithms).

The above procedure is now used in a similar fashion as for the discrete distance mapping. Euclidean discs are expanded in V from all the vertices in V_Γ and the minimum value of the radius of such discs overlapping at a vertex

u is taken as $\mathrm{DT}_\mathrm{E}^2(u)$. The major difference is that distance map values are updated only when the complete disc of radius R has been expanded. This is equivalent to superimposing the newly calculated distance values at vertices with respect to the current vertex $s \in V_\Gamma$ on the temporary Euclidean distance map calculated at the previous iteration and taking the minimum value at each pixel. Such a restriction allows for emulating parallel processing and guarantees the correctness of the resulting Euclidean distance map.

An example implementation of this algorithm is given in Algorithm 5.7. Note that, at this stage, the grid graph is not complete. However, by definition of the Euclidean distance label, the shortest path from s to p will never be constrained by the border Γ, so that Euclidean distance values deduced from the pair of integer values $(k_\mathrm{a}(s,p), k_\mathrm{b}(s,p))$ remain correct. Before expanding a Euclidean disc from a vertex s, temporary labels $d(u)$ should be reinitialised to either 0 or infinity. This can be done on-line when expanding the Euclidean disc by associating a flag with a label $d(u)$ representing the number of the iteration at which $d(u)$ has last been modified [90]. Such a flag avoids the complete scan of V for re-initialisation and therefore decreases the complexity of this procedure.

```
[——— Initialisation ———]
for all vertices q ∈ V do
  d_glob(q) ← +∞
d_glob(s) ← 0; k_a(s, s) ← 0; k_b(s, s) ← 0
[——— Main procedure ———]
for all vertices s in V_Γ do
  Initialise the search list Λ
  Initialise labels d(.)
  Characterise a Euclidean disc centred at s of radius R using labels d(u).
  for all vertices u in this disc do
    if d_glob(u) > d(u) then        [ Stores the final Euclidean distance map ]
      d_glob(u) ← d(u)
[——— Euclidean distance map characterisation ———]
for all vertices u ∈ V do
  DT_E(u)^2 ← d_glob(u)
```

Algorithm 5.7: Graph-theoretic Euclidean distance mapping

The complexity of the graph-theoretic Euclidean mapping algorithm is calculated in the same way as for the case of discrete distances. Typically, this procedure has a complexity which is expressed as $\mathcal{O}(|\Gamma|.R^2 \log R^2)$. Following Remark 5.10, it is important to note that in binary line images the value of R is typically small, so that this complexity can compare well with the optimal complexity $\mathcal{O}(W \times H)$.

In the previous algorithm, a value corresponding to $\mathrm{DT_E}(u)^2$ is obtained at each vertex u of the grid graph. A simple modification of this algorithm would allow for storing $k_{\mathrm{a}}(s,u)$ and $k_{\mathrm{b}}(s,u)$ at each vertex in order to facilitate the storage and avoid large integer values due to the square of Euclidean distances. Moreover, the storage of a variable $\mathrm{ref}(u)$ at each vertex u that points to its closest vertex s in V_Γ would store the same information as in a signed Euclidean distance map (e.g. see [176]). Finally, we emphasise the fact that, when using the Euclidean distance, the particular properties of the graph-theoretic approach given in Proposition 5.11 remain valid.

5.3.3 Other approaches

In this section, we review other efficient Euclidean distance mapping algorithms which have been proposed in the literature. Typically, for a square binary digital image image of size $W \times W$, a sequential algorithm runs optimally in $\mathcal{O}(W^2)$ operations and a parallel algorithm runs optimally in $\mathcal{O}(W)$ operations with W^2 processors.

In [72], the authors proposed an $\mathcal{O}(W^2 \log W)$ sequential algorithm which generates an exact Euclidean distance map column by column in a "divide and conquer" approach. The complexity of this algorithm is reduced to $\mathcal{O}(W^2)$ by dividing alternately lines and columns [24]. We also point out a Euclidean distance mapping technique based on the partial construction of Voronoi diagrams in the image which leads also to a complexity of $\mathcal{O}(W^2)$ on a sequential architecture [18].

We review here an original method proposed in [46, 63] which leads to a $\mathcal{O}(W^2)$ sequential algorithm whose parallel implementation in $\mathcal{O}(\frac{W^2}{N_{\mathrm{p}}})$ on N_{p} processors is shown to be particularly efficient.

The basic idea of this method is that the value of the Euclidean distance transformation at a pixel $p = (x_p, y_p)$ can be expressed as

$$\mathrm{DT_E}(p) = \sqrt{(x_p - x_q)^2 + (y_q - y_s)^2}$$

where $s = (x_s, y_s)$ is pixel in Γ closest to p and $q = (x_q, y_q)$ is the pixel such that $x_q = x_s$ and $y_q = y_p$. Such a decomposition is illustrated in Figure 5.14(A), where $p = (i, j)$ and $s = (k, l)$.

Based on this decomposition, the Euclidean distance mapping algorithm is as follows. A temporary map $(g_{ij})_{i,j}$ is defined such that

$$g_{ij} = \min_{0 \leq k \leq W-1} \{|j - k| \text{ such that } \mathrm{pixel}(i,k) \in \Gamma\}$$

and $g_{ij} = +\infty$, for all j in a column i which does not contain any pixel of Γ. An example of this map is shown in Figure 5.14(B), where the original image is that presented in Figure 5.3(A) and crosses represent infinite values.

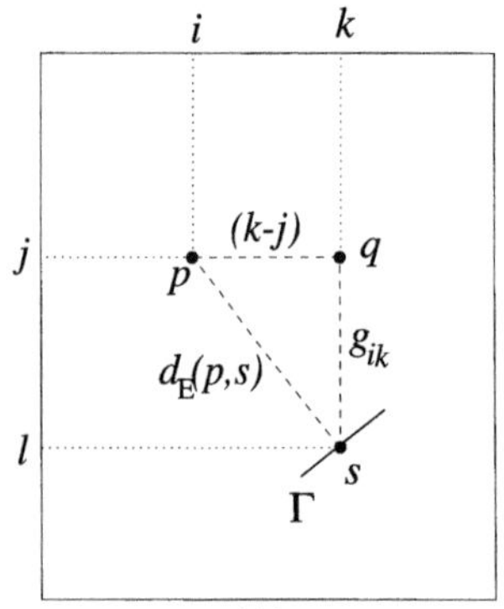

(A)

×	×	×	3	×	×	×
×	×	×	2	×	×	×
×	×	×	1	×	×	×
×	×	×	0	×	×	×
×	×	×	0	×	×	×
×	×	×	1	×	×	×
×	×	×	2	×	×	×
×	×	×	3	×	×	×

(B)

18	13	10	9	10	13	18
13	8	5	4	5	8	13
10	5	2	1	2	5	10
9	4	1	0	1	4	9
9	4	1	0	1	4	9
10	5	2	1	2	5	10
13	8	5	4	5	8	13
18	13	10	9	10	13	18

(C)

Figure 5.14 (A) Decomposition for the calculation of the Euclidean distance. (B) Temporary map $(g_{ij})_{i,j}$. (C) Final Euclidean distance map

On each row j, the following equation is then valid:

$$p = (i, j) \Rightarrow \mathrm{DT}_{\mathrm{E}}^2(p) = \min_{0 \le k \le W-1} \{(i-k)^2 + g_{kj}^2\}$$

A basic algorithm whose complexity is $\mathcal{O}(W^3)$ can readily be derived from these equations. An algorithm based on the calculation of the lower envelope of the family of functions $(f_i(k) = (i-k)^2 + g_{kj}^2)_i$ is proposed in [63]. This implementation also leads to an optimal complexity of $\mathcal{O}(W^2)$ on sequential architecture.

For the sake of completeness, we point out a technique based on greyscale mathematical morphology for the computation of Euclidean distance maps proposed in [157, 158] and completed in [130]. Typically, morphological filters corresponding to distance masks are applied over the image in a similar fashion as in Sections 5.2.1 and 5.3.1. This approach is outside the scope of this book and is not detailed further here. A similar approach that can be applied to different lattices is also presented in [60]. Finally, distance mapping applied to grey level images is considered in [178] and an extension of Euclidean distance mapping to higher dimensions is detailed in [129].

5.4 Related results

Further results concerning distance transformations and distance maps are reviewed here. Firstly, the constrained distance mapping problem is introduced and references in which the reader will find solutions detailed in different particular contexts are summarised. Similarly, Section 5.4.2 creates a link between Voronoi diagrams and distance maps. Again, references concerned with the study of this relationship are given and summarised. Finally, we give some examples for the representation of distance maps and show how these representations can be used to analyse the compatibility between discrete and Euclidean distance maps.

5.4.1 The constrained distance mapping problem

The problem of constrained distance mapping is that of calculating the length of the path from each point $p \in P$ to its closest s in $\Gamma \subseteq P$ where some subsets B_i of P form obstacles in P. This problem is illustrated in Figure 5.15, where P is the complete image, Γ is the point represented by the central white disc and obstacles B_i are represented as black polygons. Constrained shortest paths from each point p_i (shaded discs) are represented by dashed lines.

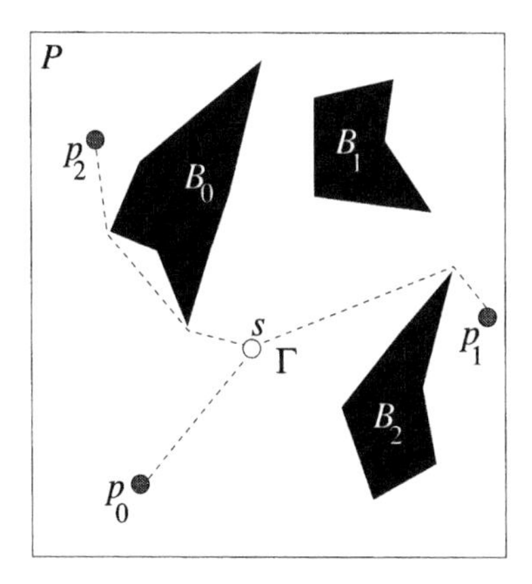

Figure 5.15 The constrained distance mapping problem

This problem finds application in different fields of computer vision such as path planning for robot vision. It has been studied as an extension of the distance mapping problem and solutions have been suggested.

More precisely, in [83], the authors consider obstacles as polygons and determine shortest paths by examining the corners of such polygons. A pixel-based adaption of this algorithm is proposed in [38], where the image is digitised and a path planning procedure is operated in this image.

In [123,124], the contour processing distance mapping is adapted to the discrete constrained case by basically not selecting obstacle pixels during a modified distance propagation process. Similar procedures for the discrete and continuous cases have also been presented in [119, 165].

It should be noted that the graph-theoretic approach presented in the previous section is readily suitable for solving the constrained distance mapping problem. In most pixel-based algorithms, problems are due to the fact that the image is scanned entirely for propagation and a pixel can represent essentially two states (i.e. $p \in \Gamma$ or $p \notin \Gamma$). Adding a third state (i.e. $p \in B_i$) causes the problem to become non-trivial. By contrast, the grid graph concept readily allows for the distance mapping of non-convex components. The grid graph is constructed by excluding obstacle pixels and the graph-theoretic distance mapping algorithms can be run in this grid graph with no modification.

5.4.2 Voronoi diagrams

As mentioned earlier, Voronoi diagrams and distance maps can be closely related. Examples are given here as to how one can deduce Voronoi diagrams from distance maps. Firstly, assuming that a point $\mathrm{ref}(p) \in \Gamma$ is stored during the construction of the distance map such that $\mathrm{DT}(p) = d(p, \mathrm{ref}(p))$, the Voronoi cell corresponding to a point $q \in \Gamma$ is the set of points p such that $\mathrm{ref}(p) = q$. Such a point $\mathrm{ref}(p)$ is readily defined in a signed distance map where the coordinates of the vector $(p, \mathrm{ref}(p))$ are stored at each point $p \in P$. Moreover, as pointed out earlier, the graph-theoretic approach for the computation of distance maps also directly allows for the storage of such a reference point $\mathrm{ref}(p)$.

In summary, given a set of points $\Gamma \subset P$ as seeds of the Voronoi diagram in P, computing the distance map of P with respect to Γ is exactly equivalent to computing the Voronoi diagram of Γ in P, and vice versa [1]. The advantage of the computation of Voronoi diagrams via distance maps is that it allows for the computation of generalised Voronoi diagrams where the set of seeds Γ is not composed of single points but rather of groups of points forming objects in P.

Examples of such computations are illustrated in Figure 5.17, where Γ is the set of pixels in Figure 5.16. Voronoi diagrams associated with this set are obtained in Figure 5.17 by associating the same grey level to all points p whose reference point $\mathrm{ref}(p)$ is a given point $q \in \Gamma$.

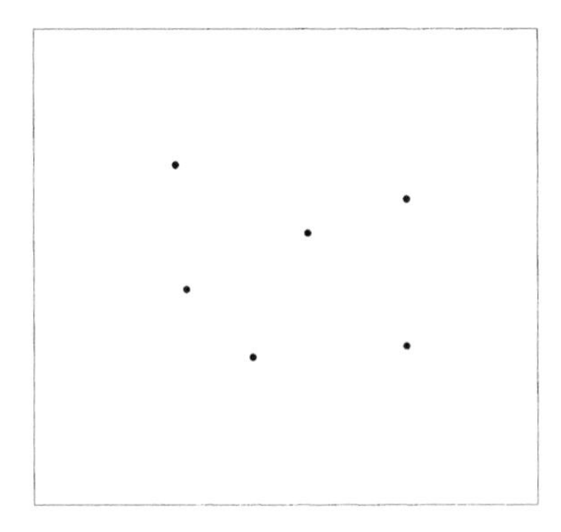

Figure 5.16 Simple pixel set for the calculation of Voronoi diagrams

Moreover, a white pixel represents a discrete point which is equidistant to two or more seeds. In this respect, it can be noted that the border of Voronoi cells may be wider that one pixel. Using this illustration, one can clearly note the difference between discrete and Euclidean Voronoi diagrams.

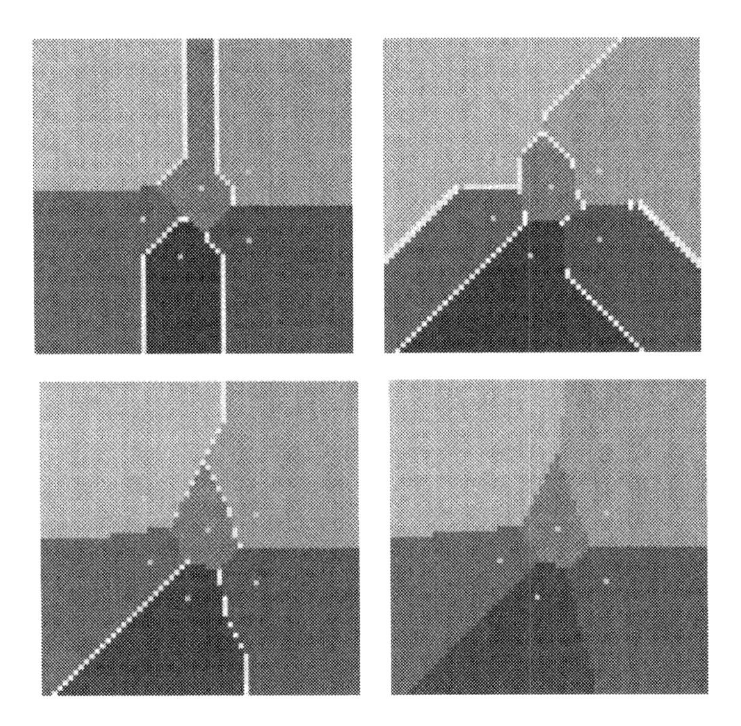

Figure 5.17 Voronoi diagrams deduced from distance maps. Upper line: Left: d_4 distance. Right: d_8 distance. Lower line: Left: $d_{3,4}$ distance. Right: Euclidean distance

5.4.3 Representation and analysis of distance maps

The distance map of an image is typically represented by a grey-level image where the grey level at a pixel represents its distance transformation value on the grey scale. Examples of such a representation are given in Figure 5.18.

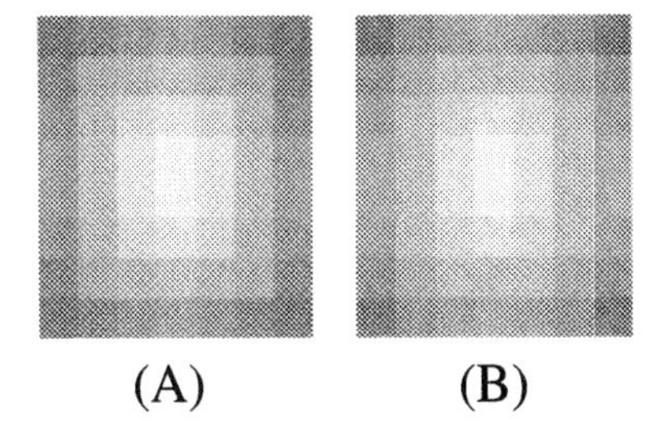

(A) (B)

Figure 5.18 Grey level representation of a distance map. (A) Discrete distance $d_{3,4}$. (B) Euclidean distance

Figure 5.18(A) shows the discrete distance map of the binary image used as an example in this chapter and presented in Figure 5.3(A). The chamfer distance $d_{3,4}$ was used for this distance mapping. The darker the grey level at a pixel, the smaller the value of distance transformation at this pixel. Figure 5.18(B) shows the Euclidean distance map of the same image whose values were presented in Figure 5.11(D). Clearly, both representations are found similar since distance values are limited in this example.

More generally, distance maps may be used to analyse discrete distances and their compatibility with the Euclidean distance. For example, Figure 5.19 illustrates distance maps of an image containing a unique central pixel as set Γ. These distance maps were obtained using discrete distances d_4, d_8, $d_{3,4}$ and Euclidean distances, respectively. Using such an image, the distance is propagated outwards from the central pixel so that the resulting distance map illustrates the shape of the unit disc corresponding to the distance in question.

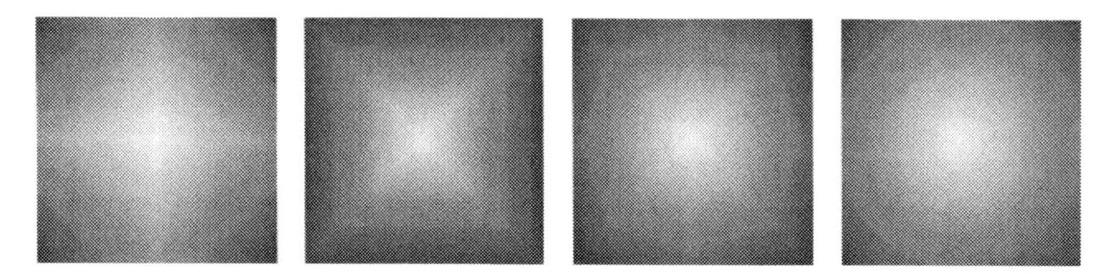

Figure 5.19 Discs highlighted using the distance map of a trivial image. From left to right, d_4, d_8, $d_{3,4}$ and Euclidean distances

Similarly, using the same representation, Figure 5.20 shows the error made between discrete and Euclidean distances. At each pixel, this error is calculated as:

- d_4: $|\mathrm{DT}_4(p) - \mathrm{DT}_\mathrm{E}(p)|$ (left).
- d_8: $|\mathrm{DT}_8(p) - \mathrm{DT}_\mathrm{E}(p)|$ (centre).
- $d_{3,4}$: $\left|\frac{\mathrm{DT}_{3,4}(p)}{3} - \mathrm{DT}_\mathrm{E}(p)\right|$ (right).

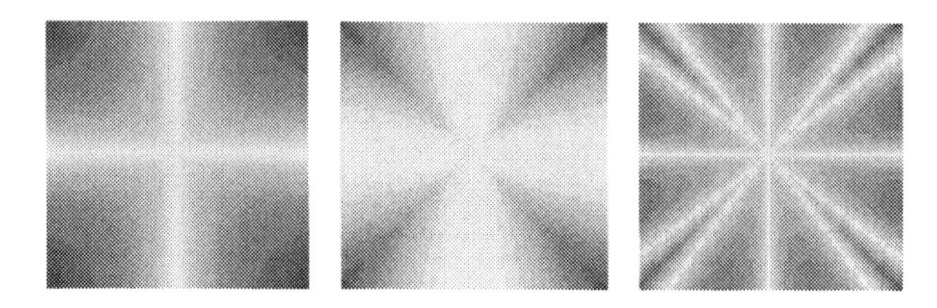

Figure 5.20 Compatibility between discrete and Euclidean distances. From left to right, d_4, d_8 and $d_{3,4}$ distances

The grey level at a pixel represents the error made between respective distances. The darker the pixel is, the higher the error is. It is clearly seen that the error is minimum along specific directions which have been pointed out in Section 1.5. Therefore, based on such representations, one can readily appreciate the quality of the approximation of Euclidean distance values by the values of a given discrete distance.

Discrete and Euclidean distance mapping will prove fundamental in the development of algorithms for binary image analysis and mostly binary *line* image analysis. Results and algorithms developed in this chapter will be fully exploited in the study of models for image representations in the following chapters.

Chapter 6

BINARY DIGITAL IMAGE CHARACTERISTICS

In the model of a binary digital image, two subsets of pixels are commonly defined within the image, namely, the foreground (i.e. black pixels) and the background (i.e. white pixels). Binary digital image analysis is essentially concerned with the characterisation of properties in the set of foreground pixels. In Chapter 2, discrete geometry was developed in order to characterise discrete equivalents for well-known properties in Euclidean geometry (e.g. straightness and convexity). This chapter introduces the morphological study of sets of discrete points representing pixels in a binary image. This work relies on the definition of connected components given in Definition 1.6, which itself depends on the definition of a neighbourhood for a pixel. Following the conclusions derived in Section 1.4.4, we will mostly concentrate on neighbourhoods defined on square lattices (see Section 1.2.3). However, it will be apparent that extensions to other lattices are straightforward in most cases. Once components are characterised, the definition of morphological factors allows for their analysis at a global level. For example, classifications of components for further recognition processing can be achieved using such factors. This chapter is organised as follows. Section 6.1 introduces the component labelling problem. After such a processing step, each connected component is treated as a separate part of the image and forms the basic entity for morphological study. However, before initiating such a study, it may be necessary to remove redundant or unwanted information from a component. Such information is referred to as noise in a general sense and, in Section 6.2, methods for reducing it are presented. Finally, Section 6.3 presents a set of factors that can be computed for representing the major information contained in a component.

6.1 Connected component labelling

The information contained in a binary digital image is mostly represented by the size, shape and location of connected components in its foreground. The notion of connectivity in the component varies depending upon the application and the context in which it is developed. Based on the study of neighbourhoods presented in Section 1.2, it is essential to derive efficient methods that identify connected components in order to process them separately for analysis. The general approach to tackling this problem is to associate with each pixel of a connected component a label by which the complete component in question is identified (e.g. the component number). Different techniques have been developed for achieving such a labelling. One common idea is to propagate component labels throughout the image. In this respect, component labelling can be compared to distance mapping presented in Chapter 5, where the propagation of a distance label forms the basis of the computation. By analogy, the component labelling process results in the *label map* which is a matrix $(\text{LABEL}(p))_p$ of the same size as the image, where each value $\text{LABEL}(p)$ indicates the component label at pixel p.

Section 6.1.1 therefore presents an approach similar to that of distance mapping using masks for computing label maps. Both sequential and parallel architectures are considered for implementation. Section 6.1.2 takes advantage of the graph-theoretic approach developed in Chapter 3 and proposes an efficient technique that achieves component labelling using spanning trees. Both sections describe algorithms and include a study of complexities.

6.1.1 Matrix-based approach

Component label propagation using the concept of masks operating on a matrix-form of the image is reviewed in this section. Typically, a mask corresponding to the neighbourhood in question is defined and is swept over the image in a sequence similar to that used for distance mapping (see Section 5.2). The major difference is that more than two passes (in the sequential case) may be needed for the complete update of the label map.

In a general context, this procedure can be described as follows. Given an binary image and F the set of foreground pixels in the image, the aim is to associate a component label $\text{LABEL}(p)$ to each pixel p in the image such that:

(i) $\text{LABEL}(p) = 0$ if $p \in F^{\mathsf{c}}$ and $\text{LABEL}(p) = +\infty$ if $p \in F$

(ii) $\text{LABEL}(p) = \text{LABEL}(q)$ if and only if pixels $p \in F$ and $q \in F$ are in the same connected component. Otherwise, $\text{LABEL}(p) \neq \text{LABEL}(q)$.

The label map $(\text{LABEL}(p))_p^{(0)}$ is first initialised using Equation (6.1) below.

$$\begin{cases} \text{LABEL}^{(0)}(p) = +\infty & \forall p \in F \\ \text{LABEL}^{(0)}(p) = 0 & \forall p \in F^{\mathsf{c}} \end{cases} \tag{6.1}$$

Then, at iteration $t > 0$, values in the label map $(\text{LABEL}(p))_p^{(t)}$ are updated using Equation (6.2).

$$\text{LABEL}^{(t)}(p) = \min\left\{\text{LABEL}^{(t-1)}(p)\ ;\ \lambda\ ;\ L_{\min}(p)\right\} \tag{6.2}$$

where

$$L_{\min}(p) = \min_{kl}\{\text{LABEL}^{(t-1)}(q)\ /\ q = (x_p + k, y_p + k) \in F\}$$

and parameter $\lambda > 0$ is a component counter and is incremented by 1 any time the value of $\text{LABEL}^{(t)}(p)$ is set to λ. Indices k and l define the neighbourhood for the connectivity in question. In other words, when possible, a label LABEL(p) at a pixel p is decreased to the value LABEL(q), the label of a neighbouring foreground pixel $q = (x_p + k, y_p + l)$. In the case where such an update is not possible and LABEL(p) is not finite, pixel p is assumed to belong to a new component and is therefore labelled λ. Clearly, the value of parameter λ is always greater by 1 than the number of components currently detected. This process is repeated until no change is made in the label map during the current iteration. At the end of the component labelling process, pixels p associated with labels $\text{LABEL}(p) = 0$ belong to the background F^c. All pixels p in a given connected component of F are associated with the same label $\text{LABEL}(p) > 0$.

The concept of masks is embedded in this approach in the sense that the sequence in which pixels are updated and neighbouring pixels considered at each iteration match the criteria which defined the implementation of distance mapping using masks. The implementation of this technique on both sequential and parallel architectures is now reviewed.

6.1.1.1 Sequential implementation

On sequential architectures, the basic implementation adopts exactly the general framework described above. The neighbourhood used to update the label of a given pixel is divided in two parts, thus defining two sub-masks. Examples of such masks are given in Figures 6.1(A) and (B) where starred pixels are considered as neighbours for the pixel in question, which is highlighted as a shaded box (i.e. such masks define the integers k and l in Equation (6.2)).

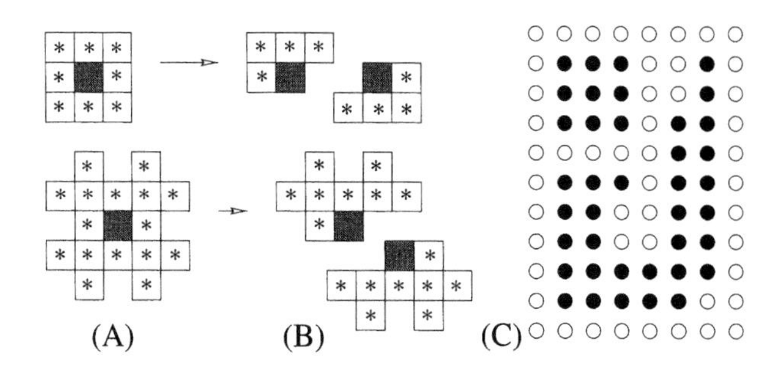

Figure 6.1 (A) Masks. (B) Resulting sub-masks. (C) Binary image

The label map is then initialised as given in Equation (6.1) and pixels are repeatedly investigated during two alternative forward and backward passes. The updating rule given by Equation (6.2) is used until a "steady state" of the label map is reached. This process can now be illustrated via the following example.

Example: Sequential connected component labelling

Consider the image presented in Figure 6.1(C). Clearly, using 8-connectivity, two components are to be detected in the image. The label map is first initialised

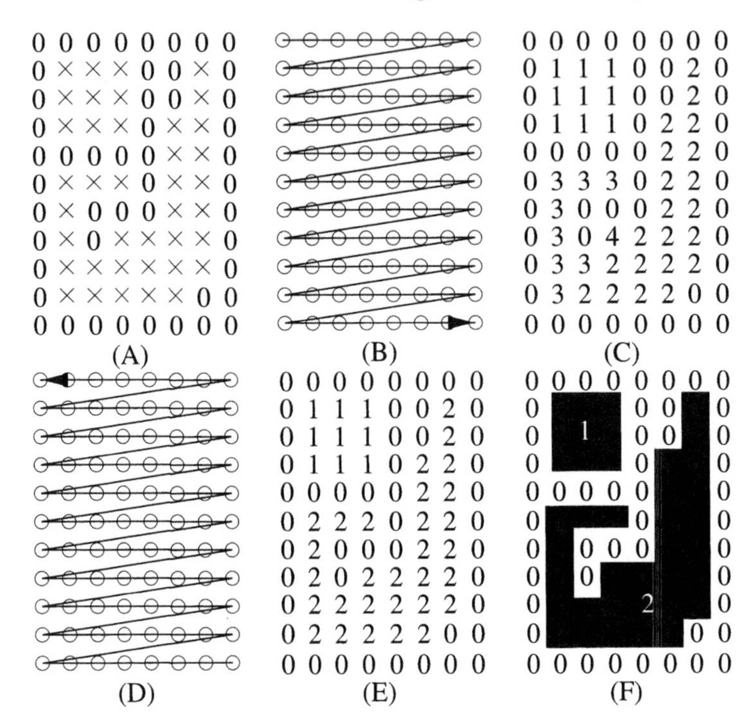

Figure 6.2 Sequential connected component labelling

using Equation (6.1), as illustrated in Figure 6.2(A), where crosses represent infinite values. The first sub-mask corresponding to the upper part of the neighbourhood is positioned at each pixel in the image, following the sequence shown Figure 6.2(B) and the updating rule in Equation (6.2) is operated. This pass results in the temporary label map shown in Figure 6.2(C). Then, the second sub-mask is swept over the image, following the revert sequence (Figure 6.2(D)) and a second temporary label map is obtained (Figure 6.2(E)). The process is repeated until no change is made in the label map and results in the final label map which can be represented as in Figure 6.2(F).

Remark 6.1:

The cases of other neighbourhoods (e.g. N_4 and N_{16} on the square lattice and N_6 on the triangular lattice) are easily adapted from the above example.

From the previous example, it is clear that the number of passes needed for the complete update of the label map depends on the form of the components in the image under study. As pointed out in [22], the worst case complexity of this procedure occurs for a spiral such as that shown in Figure 6.3. In this case, the complexity is $\mathcal{O}((W.H)^2)$ for an image of size $W \times H$.

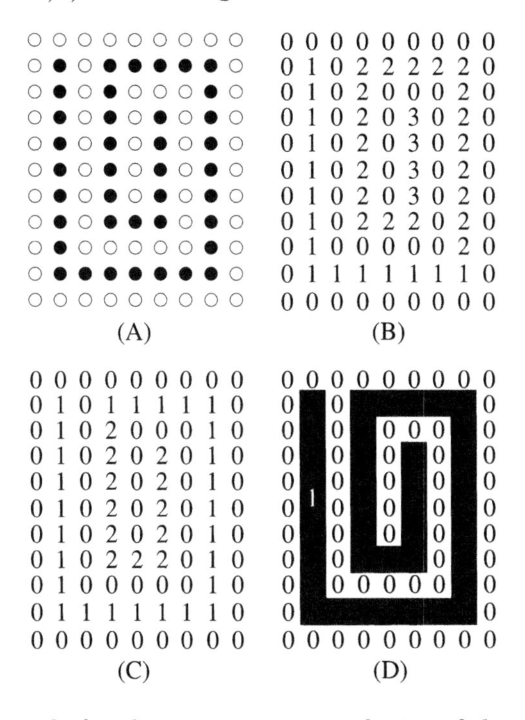

Figure 6.3 An example for the worst case complexity of the sequential labelling algorithm

In order to reduce this complexity, improved sequential procedures have been proposed. Such procedures make use of look-up tables which create relationships between temporary labels during a pass [141]. From Figure 6.2(C), it can readily be noted that labels "3" and "4" are equivalent to label "2", since both are contained together in the neighbourhood of some pixel. Therefore, by storing this information in a table, labels can be updated before initiating the next pass. In this case, the total number of passes may be reduced and the worst case complexity decreased.

Other more elaborate sequential component labelling techniques exist, based on horizontal sections created by consecutive foreground pixels in a line. Such techniques allow for storing more relationships between temporary labels, therefore decreasing further the total number of passes necessary for the complete labelling process. However, these procedures may require the management of large look-up tables, which defeats the simplicity of the approach.

6.1.1.2 Parallel implementation

The parallel implementation of the label propagation technique can be summarised as follows. The label map is initialised by associating a different label to each pixel in the foreground of the image. Then, during each parallel iteration, the label at every foreground pixel is updated according to the minimum value of labels of foreground pixels in the (complete) neighbourhood of the pixel in question. Components labels are therefore propagated along one move at each iteration. This process results in a label map where each component is labelled with the minimum value of all initial labels associated with pixels in that component. Labels can then be updated in a consecutive sequence using a look-up table in a post-processing iteration. This procedure as applied to the binary image presented in Figure 6.1(C) is illustrated in Figure 6.4.

$$
\begin{array}{cccccccc}
0&0&0&0&0&0&0&0\\
0&1&2&3&0&0&4&0\\
0&5&6&7&0&0&8&0\\
0&9&10&11&0&12&13&0\\
0&0&0&0&0&14&15&0\\
0&16&17&18&0&19&20&0\\
0&21&0&0&0&22&23&0\\
0&24&0&25&27&28&29&0\\
0&30&31&32&33&34&35&0\\
0&36&37&38&39&40&0&0\\
0&0&0&0&0&0&0&0
\end{array}
\qquad
\begin{array}{cccccccc}
0&0&0&0&0&0&0&0\\
0&1&1&1&0&0&4&0\\
0&1&1&1&0&0&4&0\\
0&1&1&1&0&4&4&0\\
0&0&0&0&0&4&4&0\\
0&16&16&16&0&8&8&0\\
0&16&0&0&0&12&12&0\\
0&16&0&19&14&14&14&0\\
0&16&16&19&19&19&19&0\\
0&21&21&21&22&22&0&0\\
0&0&0&0&0&0&0&0
\end{array}
\qquad
\begin{array}{cccccccc}
0&0&0&0&0&0&0&0\\
0&1&1&1&0&0&4&0\\
0&1&1&1&0&0&4&0\\
0&1&1&1&0&4&4&0\\
0&0&0&0&0&4&4&0\\
0&4&4&4&0&4&4&0\\
0&4&0&0&0&4&4&0\\
0&4&0&4&4&4&4&0\\
0&4&4&4&4&4&4&0\\
0&4&4&4&4&4&0&0\\
0&0&0&0&0&0&0&0
\end{array}
$$

(A) (B) (C)

Figure 6.4 Parallel connected component labelling

Figure 6.4(A) shows the initial label map $(\text{LABEL}(p))_p^{(0)}$ and Figure 6.4(B) illustrates a temporary label map obtained after three iterations (i.e. the map $(\text{LABEL}(p))_p^{(3)}$). The final label map (before post-processing) is presented in Figure 6.4(C). Label "4" can then be replaced by "2" for obtaining the same result as in Figure 6.2(F).

Clearly, the number of parallel iterations needed for the complete update of the label map is bounded by the maximum cardinality n_{max} of a shortest path in a component. The worst case complexity in this case is therefore again obtained for elongated components such as a spiral or, more generally, space-filling curves (e.g. Hilbert curve). In the example presented in Figure 6.4, $n_{\text{max}} = 11$ and all 11 parallel iterations are needed for obtaining the final label map.

A number of parallel component labelling algorithms have been described which are dedicated to specific architectures. For more detail on these procedures, the interest reader may consult [6, 27, 110].

6.1.2 Graph-theoretic approach

Algorithmic graph theory introduced in Chapter 3 provides efficient tools for the solution of the component labelling problem. The basis of graph-theoretic component labelling is the characterisation of a spanning forest in the foreground grid graph. In other words, a component will be completely labelled by expanding a tree that spans all the vertices contained in its grid graph. Shortest path spanning tree procedures are selected for this purpose. More formally, graph-theoretic component labelling can be described as follows.

The grid graph $G = (V, A)$ including all foreground pixels as vertices is assumed to be constructed for the neighbourhood in question. The label map is initialised by associating two different labels to background and foreground vertices, respectively (e.g. using Equation (6.1)). The set V is scanned and, each time a vertex s whose label corresponds to the initial foreground label is met, it becomes the root of a shortest path spanning tree. Arc lengths are set to unity and Dial's shortest path search procedure is adapted for this purpose as follows. During the shortest path search, each time a vertex is included in the spanning tree, its label is set to the value of a component counter $\lambda > 0$. As result, all vertices spanned by this tree are labelled with the current value of λ, thus defining a component. The counter λ is incremented and the vertex scan is then resumed from the vertex next to s. The procedure terminates when all vertices are associated with a non-zero finite label (i.e. when all vertices have been scanned).

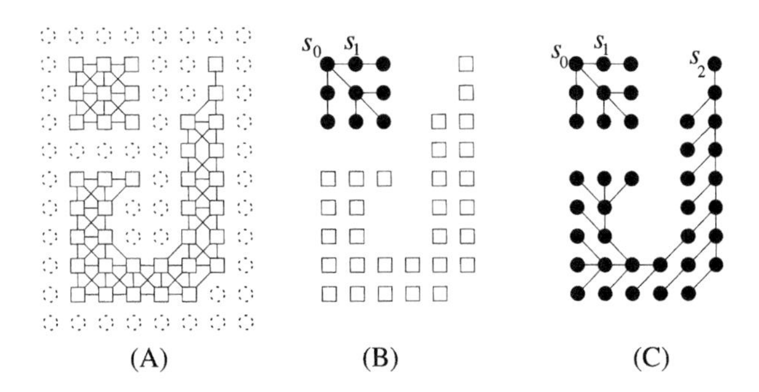

Figure 6.5 Graph-theoretic connected component labelling

Figure 6.5 illustrates the graph-theoretic approach applied to component labelling. The binary image presented in Figure 6.1(C) is used again. The 8-grid graph $G = (V, A)$ is shown in Figure 6.5(A), where background pixels are represented as dotted circles to emphasise the fact that they are not mapped in G. Non-labelled vertices are displayed as squares ($\square$), whereas vertices associated with a component label are shown as black discs ($\bullet$). The set of vertices V is then investigated sequentially and s_0 is met as a vertex whose label is equal to the original foreground label. Dial's procedure is therefore initiated from s_0 and results in the shortest path spanning tree shown in Figure 6.5(B). The grid graph is not shown in Figures 6.5(B) and (C) for clarity. All vertices in this tree are then labelled with the value of the component counter $\lambda = 1$. This counter is incremented and the vertex scan restarts at vertex s_1 and vertex s_2 is met as another vertex associated with the original foreground label. The shortest path search procedure is used again and results in a second spanning tree, based on the vertices of a separate component (Figure 6.5(C)). At this stage, the rest of the vertices are scanned and, since all vertices are included in either of the two spanning trees, their labels have been updated during the tree construction and the procedure terminates.

Since arc lengths are set to unity, Dial's shortest path search procedure proves efficient in this case and leads to an overall complexity of $\mathcal{O}(W.H)$ for an image of size $W \times H$. It is important to emphasise the fact that, since only foreground pixels are to be mapped onto vertices in G, after the grid graph construction (i.e. $\mathcal{O}(W.H)$ procedure), the complexity of the labelling procedure does not depend on the form of connected components. Moreover, it is easily seen that global information such as the number of pixels in a component can readily be accessed by this approach (see also Section 6.3). By contrast, other approaches require a final scan for obtaining such information.

6.2 Noise reduction

Binary images considered for analysis generally result from the digitisation of some continuous images. Similarly, such binary images can be created by thresholding the grey-level at each pixel in grey scale images. This processing represents the most basic operation in the class of *segmentation processes*. Global or local adaptive thresholds can be considered. For an extended account on these techniques, the reader is referred to the relevant literature (e.g. [62]). Depending on parameters such as resolution and grey-level threshold, the binary image may be altered by noise. In this section, noise refers to either black or white pixels added randomly to the image (i.e. salt and pepper noise) or a group of (black or white) pixels added to the image. The first type of noise typically arises in an acquisition process, whereas the second type of noise typically results from an inaccurate thresholding process.

In both cases, this noise is to be removed from the binary digital image for accurate analysis. Noise removal is based on the knowledge of the type of features present in the image. For example, it can be required that all connected components are discrete convex. Different approaches are possible and are briefly reviewed here. The first approach presented in Section 6.2.1 uses the concept of discrete masks introduced in Section 5.2.1 to represent local properties of digital images. The second approach consists of smoothing the contour of the component, in order to satisfy a convexity criterion or, more generally, a regularity criterion. Section 6.2.2 reviews different techniques for achieving contour smoothing. Finally, Section 6.2.3 presents a noise-removal technique based on prior knowledge of the image content. Components are divided in convex parts or, more generally, in sub-components that match a certain shape criterion. In this respect, shape factors which will be presented in Section 6.3 may be used to define the segmentation criterion.

6.2.1 Median filtering

Median filtering is known to perform well on grey-level images (including the particular case of binary images) for removing "salt and pepper" noise [62]. It can also be used for binary images where the noise to be removed consists of unwanted black and white isolated pixels. We first introduce median filtering in a general context (i.e. for grey-level or binary images). We will then present examples which highlight the performance of this technique.

A neighbourhood is first to be defined. That is, using a discrete distance function d_{D} (see Section 1.4), a discrete mask $M(p)$ located at pixel p covers all pixels q such that $d_{\mathrm{D}}(p, q) \leq D$, where D is a given discrete distance value. For example, if $d_{\mathrm{D}} = d_8$ and $D = 1$, then $M(p) = \{q \text{ such that } d_8(p, q) \leq 1\}$ is the (3×3) mask centred at p.

Median filtering now simply consists in replacing the grey-level value at any pixel p by the median of the set of grey-level values of pixels contained in the mask $M(p)$. We recall that, given an ordered set S of N values, median(S) is the middle value in S. More precisely, if $S = \{\alpha_1, \ldots, \alpha_n\}$ and $\alpha_i \geq \alpha_j$ for $i > j$, median$(S) = \alpha_{\frac{N+1}{2}}$ if N is odd and median$(S) = \frac{1}{2}(\alpha_{\frac{N}{2}} + \alpha_{\frac{N}{2}+1})$ otherwise. In the latter case, integer values can be obtained using some rounding function (e.g. see Definition 1.21).

Example: Median filtering

Consider the original digital image of a letter "e" presented in Figure 6.6(A).

Figure 6.6(B) shows the same image where uniform salt and pepper noise has been added. Results obtained by filtering this image using median masks of size (3×3), (5×5) and (7×7) are presented in Figures 6.6(C) to (E), respectively. This example highlights the need for an accurate choice of the mask size. If the chosen mask size is too large, median filtering is unable to distinguish between small components and noise. From this example, (3×3) seems to be an accurate choice. Continuing with this example, Figure 6.6(F) shows the original image filtered using a (3×3) median mask. Irregularities on the border are smoothed away.

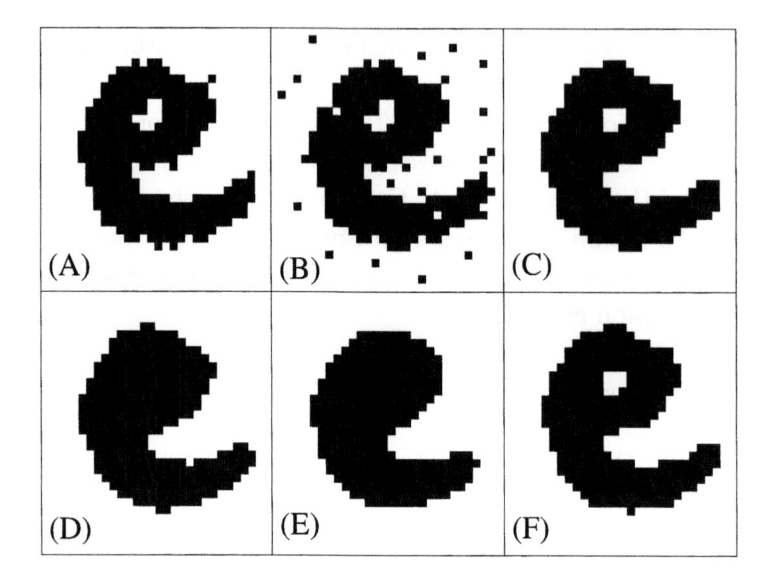

Figure 6.6 Median filtering

Sorting of the values within the mask is required prior to median value selection. Therefore, if $N_{\mathrm{M}} = |M(p)|$, the median value can be characterised in $\mathcal{O}(N_{\mathrm{M}} \log(N_{\mathrm{M}}))$ operations. However, since the mask is shifted from pixel to pixel in a similar fashion to that described in Section 5.2.1, two masks corresponding to two neighbouring pixels differ by N_{M} values only. In this case, the median value can be characterised in $\mathcal{O}(N_{\mathrm{M}})$ operations at each pixel. In any case, this operation can be considered as constant time (i.e. it does not depend on the number N_{pixels} of pixels in the image), so that the overall complexity of median filtering is $\mathcal{O}(N_{\mathrm{pixels}})$.

6.2.2 Contour smoothing

The aim here is to remove irregularities that can be encountered when following the contour of a connected component. Methods based on contour following rely on the definition of an orientation within the chosen neighbourhood. Starting from a given pixel p on the border of the component, the next pixel in the contour is found as the first neighbour of p in the neighbourhood sequence (e.g. see [22, 122]). Clearly, the contour of a connected component F is always included in its border set Γ. Contour following results in a chain-code of the contour on which the smoothing is operated.

In [179], smoothing is based on the replacement of sequences of chain-codes by other simpler sequences. Instances of such templates are given in Figure 6.7, where the original 8-chain-code sequence shown as dashed arrows between pixels p and q is to be replaced by the sequence of 8-chain-codes shown as continuous arrows.

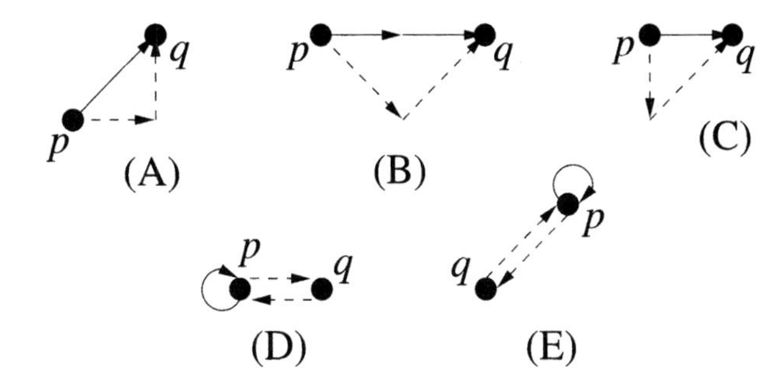

Figure 6.7 Templates for smoothing based on the chain-code of the contour

Schematically, chain-code sequences are replaced by a shorter equivalent. In Figures 6.7(D) and (E), pixel q is simply dismissed since it corresponds to a "peak" point of the contour. The smoothed contour is then to be filled using a technique such as that presented in [107]. Results of such a smoothing procedure are illustrated in Figure 6.8. Figure 6.8(A) shows the border of the original connected component used in Figure 6.6. Contour pixels are represented as black discs and interior pixels are not shown for clarity. The resulting smoothed component is shown in Figure 6.8(B), where empty circles illustrate original contour pixels that do not appear in the final smoothed contour.

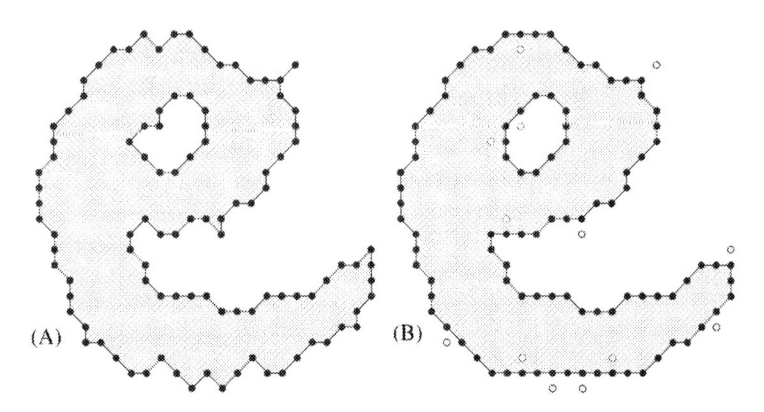

Figure 6.8 Chain-code-based smoothing

Following this approach, it should be noted that the use of shortest paths may prove efficient in smoothing contours. For example, a smoothed contour may be found as a shortest path in the grid graph including pixels within a defined neighbourhood of border pixels. A similar idea is exploited below.

Let F and F^{c} be the foreground and the background of a binary digital image, respectively. The smoothing technique presented here uses distance maps to characterise regions in the image where a smoothing operation may be performed. Distance mapping was studied in Chapter 5 and is based on the definition of a distance for the set it operates in. Let us assume the use of a (possibly discrete) distance function $d(.,.)$ in F and F^{C}. A modified distance mapping is performed over $F \cup F^{\mathsf{C}}$ (i.e. over the complete image) so that any pixel $p \in F \cup F^{\mathsf{C}}$ is associated with a label $\mathrm{DT}^{(1)}(p)$ such that

$$\mathrm{DT}^{(1)}(p) = \min_{q \in F^{\mathsf{C}}} d(p,q) \text{ if } p \in F$$

and

$$\mathrm{DT}^{(1)}(p) = \min_{q \in F} d(p,q) \text{ if } p \in F^{\mathsf{C}}$$

In other words, each pixel in the image is labelled with the minimum distance to a pixel in the complement of the set it belongs to. A set of pixels P is then defined by $P = \{p \in F \cup F^{\mathsf{C}} \text{ such that } \mathrm{DT}^{(1)}(p) \leq \tau\}$, where $\tau > 0$ is a distance threshold. Clearly, P is the set of pixels lying within a distance τ around the border separating F and F^{C}. A second distance mapping is now performed in P with respect to its border Γ_P. As result, each pixel p is labelled with a distance value $\mathrm{DT}^{(2)}(p)$ such that

$$\mathrm{DT}^{(2)}(p) = \min_{q \in \Gamma_P} d(p,q)$$

Pointers $\mathrm{ref}(p) \in \Gamma_P$ are defined as pixels that characterise the values $\mathrm{DT}^{(2)}(p)$ (i.e. $\mathrm{DT}^{(2)}(p) = d(p, \mathrm{ref}(p))$). In this context, the smoothed foreground F' is finally re-composed, following the rules:

(i) If $p \in P^{\mathsf{C}} \cap F$ then, $p \in F'$.

(ii) If $p \in P^{\mathsf{C}} \cap F^{\mathsf{C}}$ then, $p \in F'^{\mathsf{C}}$.

(iii) If $p \in P$ and $\mathrm{ref}(p) \in F$ then, $p \in F'$.

(iv) If $p \in P$ and $\mathrm{ref}(p) \in F^{\mathsf{C}}$ then, $p \in F'^{\mathsf{C}}$.

This procedure results in a smoothed contour, where isolated pixels have been merged with either F or F^{C}. Moreover, this technique involves a parameter τ which can be used for characterising the level of noise which is to be considered. A large value of τ will generate a smoothing procedure where large protrusions or dents will be removed, whereas a small value for τ may only remove pixel-size noise.

In the original [3], both distance mapping are performed via image scans using distance masks (see Section 5.2.1). A fast implementation of this technique is presented in [125] using contour processing which continues with the results presented in Section 5.3. This modification allows for considering a limited set of pixels around the border between the original sets F and F^{C}, thus improving the performance.

Figure 6.9 illustrates the smoothing process using distance mapping. A part of the noisy edge is shown in Figure 6.9(A). Since the noise corresponds to pixels added or removed in a limited area around the border, only a small value for the threshold is required for smoothing.

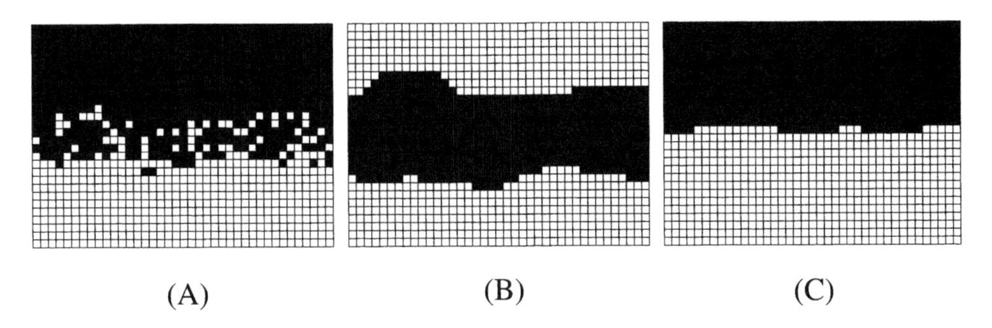

(A) (B) (C)

Figure 6.9 Contour smoothing using a distance transformation

The resulting set P is shown in Figure 6.9(B) as the set of black pixels in this binary image. The last step divides P through its medial line; the upper half is consider as foreground pixels and the lower half as background pixels. This image is then superimposed on the original set $F \cup F^{\mathrm{C}}$ and results in the smoothed contour between the sets F' and F'^{C} shown in Figure 6.9(C).

Another classic approach is based on operators from mathematical morphology. The basis of mathematical morphology is given by set theory and, more specifically, by *Minkowski algebra.* Our aim here is simply to introduce the reader to some basic morphological operators which have proved efficient in noise removal on binary images. For further details on the field of mathematical morphology and a formal introduction of the operators below, the interested reader is referred to [48,126,149,150]. For our purpose, two major morphological operations are to be defined.

Definition 6.2: Erosion and dilation operators

Given a set of pixels F and Γ its border with respect to a neighbouring relationship. For any pixel $p \in P$, $N_{\mathrm{D}}(p)$ denotes the neighbourhood of p.

(i) The dilation operator dilation(.) *applied to F results in the set*

$$\mathrm{dilation}(F) = F \cup \bigcup_{p \in F} N_{\mathrm{D}}(p)$$

(ii) The erosion operator erosion(.) *applied to the set F results in the set*

$$\mathrm{erosion}(F) = F \setminus \Gamma$$

Remark 6.3:

(i) In Minkowski algebra, dilation *and* erosion *operators are referred to as Minkowski addition (denoted $\oplus$) and subtraction (denoted $\ominus$), respectively.*

(ii) The above definition is a particular case of the dilation(.) *and* erosion(.) *classically defined using a generalised structuring element B. Here, we consider $B = N_{\mathrm{D}}$.*

Typically, the dilation and erosion operators respectively add or remove a layer of pixels on the border of F. Figure 6.10 illustrates the dilation and erosion of a generic set F. Clearly, $\mathrm{erosion}(F) \subset F \subset \mathrm{dilation}(F)$.

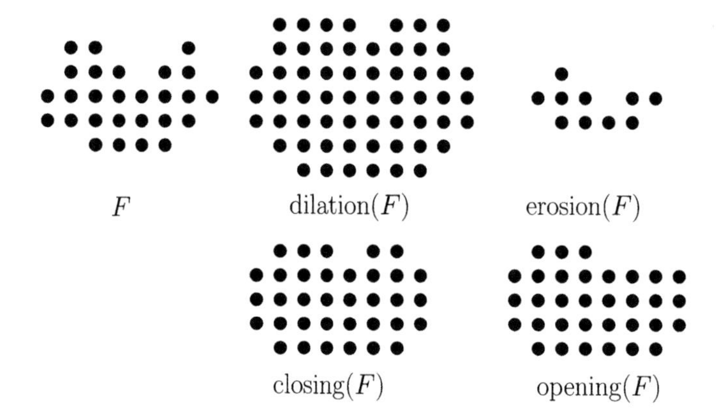

Figure 6.10 Morphological operators

By combining these basic operators, advanced morphological operators can be obtained.

Definition 6.4: Closing and opening operators

Given a set of pixels F,

(i) The closing operation applied to F results is the set closing(F)*, defined by*

$$\mathrm{closing}(F) = \mathrm{erosion}(\mathrm{dilation}(F))$$

(ii) The opening operation applied to F results is the set opening(F)*, defined by*

$$\mathrm{opening}(F) = \mathrm{dilation}(\mathrm{erosion}(F))$$

Both operations are also illustrated in Figure 6.10. Clearly, these two operators can be used for smoothing. The closing operator fills irregularities in the contour of the component, whereas the opening operator smooths these irregularities by removing them.

Clearly, if the original set F is reduced to one pixel, the set erosion(F) is the empty set $\emptyset$. Since closing($\emptyset$) $= \emptyset$, the opening operator can be used to remove spurious component in the image. Conversely, the closing operator can be used to fill spurious holes in a component. Typically, by using different combination of the erosion and dilation operators, one can obtain different levels of smoothing. For example, two-pixel sized holes will be filled using the operator erosion(erosion(dilation(dilation(.)))), and two-pixel sized components will be removed when applying the sequence dilation(dilation(erosion(erosion(.)))). In this respect, eroding the foreground of a component is seen as equivalent to dilating the background and vice versa. Therefore, a closing operation performed on F results in an equivalent image to that from an opening operation performed on F^{c} and vice versa. Hence, using combination of these operators, one can define a variety of morphological operators that allow for the characterisation of different properties in the image.

When convexity is selected as the criterion, connected components in the image are to be replaced by their respective convex hulls. Different techniques exist for the computation of the discrete convex hull of a set of discrete points. For a detailed study of discrete convexity and the details of a convex hull construction procedure, the reader is referred to Section 2.3.2.

In summary, different approaches can be taken for smoothing the contour, depending on the level of noise involved. For local smoothing, operations on chain-codes result in an accurate smooth contour. Clearly such operations cannot be performed on a noisy border such as that shown in Figure 6.9. In this case, techniques based on noise filters or techniques which operate via a threshold level for merging redundant pixels such as distance mapping smoothing or operators issued from mathematical morphology may be more efficient. Finally, based on a prior knowledge of the component, one can use geometrical properties such as convexity for achieving a smoothing which considers the global shape of the component.

6.2.3 Shape segmentation

Dividing a connected component into different parts may be needed in the case where connected components are required to satisfy a geometrical criteria. For example, when studying a blood sample image, blood cells should be separated from each other and are known to be convex. Mathematical morphology may be used in this case.

We illustrate such a process with an example. Consider the binary image shown in Figure 6.11(A). This image represents two convex sets of pixels joined with a spurious elongated object. Clearly, with the use of a suitable succession of erosion operators (three, in this case), the central object will collapse. This is highlighted by lines symbolising erosion steps in the original component. An equivalent number of dilation operations will then restore the integrity of both convex sets (Figure 6.11(B)).

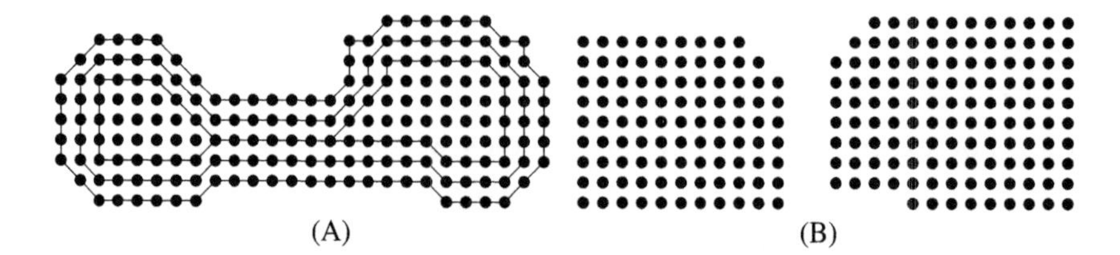

Figure 6.11 Component segmentation

However, an accurate number for such basic operations may not be known in practice, as it varies with the image and the resolution. One can design an adaptive scheme by using morphological factors described in Section 6.3.3 so that this number is automatically adapted. For example, the component may be eroded until each sub-component satisfies the convexity criterion. The decomposed figure can then be retrieved by an equivalent number of successive dilation operations.

Different approaches exist for segmenting components. In general, they rely on an iterative process that stops when a certain criterion is satisfied. This is the case, for example when using active contours (region growing) for segmenting the component into convex parts. Starting from a point in the component F, a region P is expanded into the component and convexity of P is maintained at each growing step. At some stage, P completely spans a convex sub-part of F. During the next growing step, P is forced to access a non-convex region of F. Since P is constrained by F, P itself becomes non-convex. This iteration is rejected and P is isolated as a convex sub-part of F. This process is then resumed in $F \setminus P$ until complete decomposition is achieved. Scale-space theory [84] may also be a suitable context for shape segmentation since it allows for characterising information at different levels.

6.3 Shape factors

Given a connected component, it is of interest to associate with it shape factors. Ideally, a set of shape factors associated with a set of points would enable one to uniquely characterise this component. Moreover, this set of factors should define a measure, in order to quantify similarity between the component in question with some other reference object.

In this section, different types of such factors are introduced. Generally, they can readily be associated with their continuous counterparts. Section 6.3.1 introduces topological factors which allow for a global characterisation of a component. By contrast, Section 6.3.2 defines values such as perimeter and area.

These coefficients depend on a sampling step and are referred to as geometrical factors. Finally, Section 6.3.3 presents the calculation of morphological factors which can be considered as similarity measurements against some ideal objects.

6.3.1 Topological factors

Topological factors describe a component by representing interactions between its parts. Firstly, Euler's number can be associated with a complete image containing n_{comp} different connected components. A component may contain some holes in its interior. The number of such holes in the component labelled i is denoted n^i_{hole}. In this context, Euler's number is given by the following definition.

Definition 6.5: Euler's number

Given an image containing n_{comp} different connected foreground components and assuming that each foreground component i contains n^i_{hole} holes (i.e. defines n^i_{hole} extra components in the background), Euler's number associated with the image is calculated as

$$E = n_{\text{comp}} - \sum_{i=1}^{n_{\text{comp}}} n^i_{\text{hole}}$$

In the case of discrete objects, pixels can be categorised using a connectivity factor. Two classic definitions exist for the cases of the 4- and 8-neighbourhoods on the square lattice, respectively.

Definition 6.6: Crossing and connectivity numbers

Pixels q_i $(i = 0, \ldots, 7)$ in the 8-neighbourhood of a pixel p are ordered in a clockwise fashion, starting from any 4-neighbour of p and are associated with a value $q_i = 0$ or $q_i = 1$, depending on whether q_i is a white or black pixel, respectively. Then,

(i) The crossing number $\chi_4(p)$ is defined by

$$\chi_4(p) = \prod_{i=0}^{7} q_i + \frac{1}{2}\sum_{i=0}^{7} |q_{i+1} - q_i|$$

and indicates the number of 4-connected components in the 8-neighbourhood of p.

(ii) The connectivity number $C_8(p)$ is defined by

$$C_8(p) = q_0 q_2 q_4 q_6 + \sum_{i=0}^{3} (\overline{q}_{2i} - \overline{q}_{2i}\overline{q}_{2i+1}\overline{q}_{2i+2})$$

where $\overline{q}_i = 1 - q_i$. The connectivity number indicates the number of 8-connected components in the 8-neighbourhood of p.

Index i is interpreted as modulo 8 (i.e. if $i > 7$, the value $i - 8$ is considered).

Using Definition 6.6, pixels p in a 4-connected component F are differentiated as follows.

(i) If $\chi_4(p) = 0$, then p is an isolated point (i.e. $F = \{p\}$).

(ii) If $\chi_4(p) = 1$, then p is a either a border point or an interior point.

(iii) If $\chi_4(p) = 2$, then p is essential to keep F 4-connected.

(iv) If $\chi_4(p) = 3$, then p is a branching point.

(v) If $\chi_4(p) = 4$, then p is a crossing point.

Characteristics for such points in the 4-connected set F are illustrated in Figure 6.12(A), where symbols $\circ$, $\square$, $\diamond$, $\perp$ and $\times$ represent points p where the connectivity number takes the values $\chi_4(p) = 0$, $\chi_4(p) = 1$, $\chi_4(p) = 2$, $\chi_4(p) = 3$ and $\chi_4(p) = 4$, respectively.

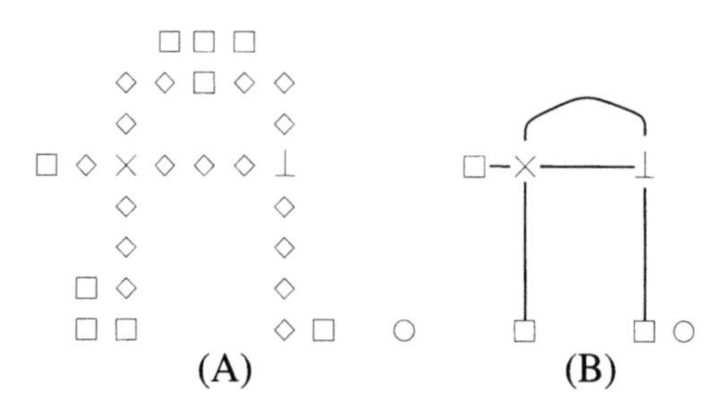

Figure 6.12 Interpretation of connectivity numbers

Such factors allow for a first estimation of a graph that describes the topology of the object (i.e. its upper-level structure). In [32], such a graph is called *graph representation* of the component. The authors of [32] point out that, since this graph is planar, Euler's formula holds for such a graph (i.e. $n_{\text{hole}} = 1 + |A| - |V|$, where V and A are the vertex and arc sets in the graph, respectively).

The graph representation highlights holes and "free ends" in the component and gives inter-relationships between parts of this objects. Clearly a wide range of shapes can be mapped onto the same graph. The graph representation is particularly meaningful when associated with line images since most of the information carried in these images is contained in their topological graph (see Section 7.3). Such graphs will therefore form an important part of the discussion in the problem of processing line images for analysis (see Chapters 7 and 8).

6.3.2 Geometrical factors

Geometrical factors associated with a shape are those which depend on a scale (i.e. where a unit is associated with their values). They include perimeter, area and values such as radii of optimal enclosed or enclosing circles. Methods for calculating them are given here and the unit considered is generally based on the size of the unit grid element (i.e. partitioning polygon). Conversion to real length units can then be done by using the value of the sampling step $h > 0$ (Definition 4.3).

6.3.2.1 Perimeter of a discrete object

The perimeter of a connected component F can be directly calculated from the chain-code of its contour. Since the chain-code is composed of unit moves, the result will clearly be an approximation of the real value of the perimeter of the preimage of F. Different estimators can be used. The simplest estimator for the perimeter of a connected component is given by the number of pixels on its contour.

Based on the study of discrete distances, the estimation of the perimeter can be refined by defining it as the sum of the lengths of the moves that compose the contour (e.g. a-, b- or c-moves). A generic scheme for defining estimators for the perimeter of a discrete object is given in [75] and is mostly based on the error made when associating lengths to moves in the contour.

6.3.2.2 Area of a polygon based on discrete points

On the square grid, unit area is chosen to be the area of the unit square. That is the square defined by the lower-left corner $p = (x_p, y_p)$ and the upper-right corner $q = (x_q, y_q) = (x_p + 1, y_p + 1)$ (see Figure 6.13(A)). The area of a discrete object is defined as the number of unit squares that it contains. Pick's formula allows for direct calculations of surfaces of polygons with discrete points as corners.

Proposition 6.7: Pick's formula [22, 148, 167]

Given a polygon $\mathcal{P}$ whose corners are discrete points, let P be the set of discrete points included in $\mathcal{P}$. Let n_B be the number of discrete points that lie exactly on the polygonal line that forms the borders of $\mathcal{P}$ and let n_I be the number of interior points (i.e. non-border points) in P. Then, $|P| = n_B + n_I$. The area of $\mathcal{P}$, denoted $\mathcal{A}(\mathcal{P})$ is the number of unit squares (possibly fractional) that are contained within $\mathcal{P}$ and is given by

$$\mathcal{A}(\mathcal{P}) = n_I + \frac{n_B}{2} - 1$$

Example: Area of a polygon based on discrete points

Consider the polygon $\mathcal{P}$ shown in Figure 6.13(B). The border of $\mathcal{P}$ is represented as a continuous bold line. Points in P are represented by discs ($\bullet$ and $\circ$). Black discs ($\bullet$) represent border points (i.e. corners and points that lie exactly on the border of $\mathcal{P}$). Empty discs ($\circ$) represent interior points. Clearly, $n_{\mathrm{I}} = 71$ and $n_{\mathrm{B}} = 10$. Therefore, $\mathcal{A}(\mathcal{P}) = n_{\mathrm{I}} + \frac{n_{\mathrm{B}}}{2} - 1 = 75$.

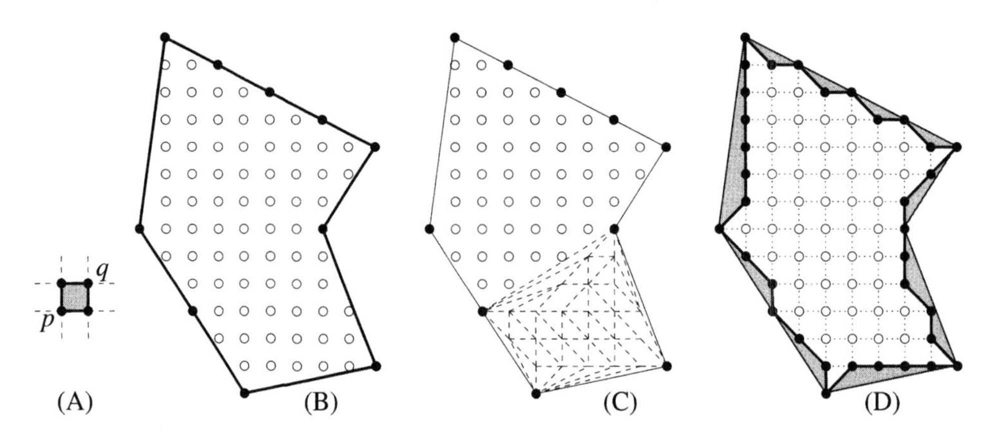

Figure 6.13 Area of a polygon. (A) Unit area. (B) Polygon $\mathcal{P}$. (C) Recursive decomposition of $\mathcal{P}$ for the proof of Proposition 6.7. (D) Polygon considered in Remark 6.8

Proof:
The proof of Proposition 6.7 relies on a recursive decomposition of the polygon $\mathcal{P}$ into sub-triangles that do not contain any discrete points in their interior or border, except at their corners (e.g. see Figure 6.13(C)). The area of such sub-triangles is constant and shown to be equal to $\frac{1}{2}$ (which corresponds to the case where $n_{\mathrm{I}} = 0$ and $n_{\mathrm{B}} = 3$). Sub-triangles are then iteratively merged and their areas summed. The count of corner points that remains at each step of the summation leads to Pick's formula. $\square$

Remark 6.8:

A contrast should be made between the area defined by $\mathcal{P}$ and the area defined by the contour of P defined from the border set Γ (see Definition 1.11). Figure 6.13(D) illustrates the resulting polygon obtained when 8-connectivity is considered in P and 4-connectivity is considered in P^{c}. The area of this polygon is clearly 63. One can verify that the difference between this and $\mathcal{A}(\mathcal{P})$ is exactly the area of the surface between the contour and $\mathcal{P}$ (shaded area in Figure 6.13(D)).

6.3.2.3 Centre, radius and diameter of a discrete object

Radius and diameter of a connected set of pixels refer to distances within this component. They can be derived directly from their graph-theoretic counterparts defined in Definition 3.12.

Definition 6.9: Centre, eccentricity, radius and diameter of a connected component

Given a connected set of pixels F and a distance function $d(.,.)$ between points of F,

(i) The eccentricity ecc(p) *of a point p in F is the maximum of $d(p,q)$ for all points $q \in F$,*

$$\mathrm{ecc}(p) = \max_{q \in F} d(p,q)$$

(ii) The centre of F is the set of points p of least eccentricity in F.

(iii) The radius of F is the value of the least eccentricity in F.

(iv) The diameter of F is the value of greatest eccentricity in F.

It is generally the case that all the above parameters are characterised using a discrete distance function in F (e.g. chamfer distance). In this case, the distance between two points in the component is considered as the length of the shortest path between these two points *within* F. Clearly, this shortest path may be constrained by the border of F. In this context, clearly, algorithmic graph theory introduced in Chapter 3 is well-suited to compute these values.

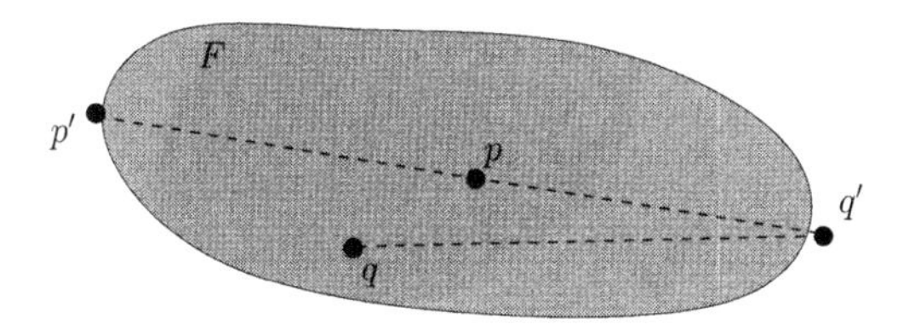

Figure 6.14 Centre, eccentricity, radius and diameter of a component

Figure 6.14 illustrates these parameters in a connected component F. Point p is the centre of F, $d(p,p')$ is the radius of F and $d(p',q')$ is the diameter of F. In this context, the value of the eccentricity at q is $\mathrm{ecc}(q) = d(q,q')$. Dashed lines represent shortest paths between respective points, defining the value of the distance function $d(.,.)$.

Similar shape factors can be deduced from the distance map of F. For example, the value $\max\{\mathrm{DT}(p)\ ;\ p \in F\}$ gives the radius of the biggest enclosed circle in F. Operations on values in the distance map such as average may also provide further shape descriptors. However, as pointed out in Chapter 5, distance maps will mostly be used for the representation of binary digital images. Their capabilities will therefore be discussed extensively in Chapter 7.

6.3.3 Morphological factors

Morphological factors refer to values or characterisations that do not involve units. They are generally calculated as a ratio between different geometrical factors of the component in question and are designed to take precise values for well-shaped components (e.g. circle and rectangle).

6.3.3.1 Centroid and moments

Centroid and moments are given as the equivalent of their continuous counterparts. Given F, a set of n pixels $p_i = (x_i, y_i)$ $(i = 1, \dots, n)$, the coordinates (x_q, y_q) of the centroid q of F are calculated by

$$x_q = \frac{1}{n}\sum_{i=1}^{n} x_i \qquad \text{and} \qquad y_q = \frac{1}{n}\sum_{i=1}^{n} y_i$$

For example, given the connected component in Figure 6.15, the highlighted pixel represents the closest pixel to its centroid.

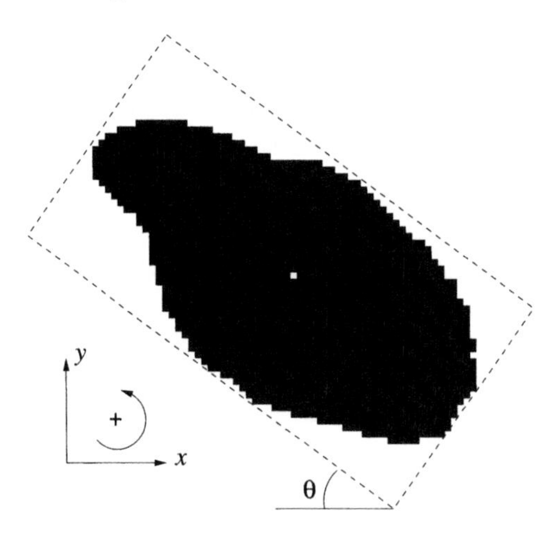

Figure 6.15 Centroid and optimal orientation of a generic component

Moments allow for a unique characterisation of a shape. Given F, a set of n pixels $p_i = (x_i, y_i)$ $(i = 1, \dots, n)$, and $q = (x_q, y_q)$ its centroid, the definition of the discrete (k, l)-order central moment $\mu_{k,l}$ of the set F is given by

$$\mu_{k,l} = \sum_{i=1}^{n} (x_i - x_q)^k (y_i - y_q)^l$$

A shape is uniquely represented by the set of all its (k, l)-order central moments $(k, l \in \mathbb{N})$. Clearly, if the shape is symmetric with respect to the diagonal axis, $\mu_{k,l} = \mu_{l,k}$ for all $k \in \mathbb{N}$ and $l \in \mathbb{N}$. Moreover, moments allow for the definition of an orientation of the component in question. Orientation is defined as an angle θ representing the angle made between the axis and the axis of least *moment of inertia.* The value of θ is obtained by the following formula.

$$\theta = \frac{1}{2} \arctan \left(\frac{2\mu_{1,1}}{\mu_{2,0} - \mu_{0,2}} \right)$$

Then, using the value of θ as angle for a rotation, the best enclosing rectangle can readily be found. In Figure 6.15, θ and the enclosing rectangle deduced from this value are illustrated.

In practical applications such as character recognition, the value of θ can be used to correct the skewness of a line in a sampled document. For example, the word "TEXT" shown in Figure 6.16 has been rotated by an angle of 32°, counterclockwise. The computation of moments and orientation gives a value $\theta \simeq 31.269055$, which is an accurate estimation of the initial value, considering that this set of point is not homogeneous.

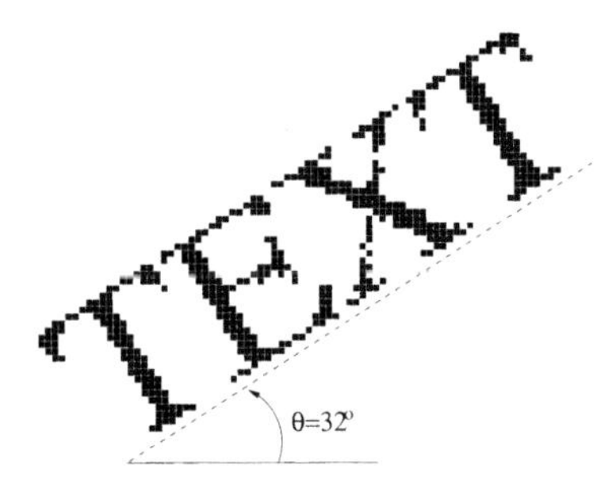

Figure 6.16 Correcting the skewness of digitised text

6.3.3.2 Compactness

The compactness of a component refers to a measure between the component and an ideal discrete object such as a rectangle or a circle.

A trivial measure is given by the saturation of the component within its enclosing rectangle. Two different components are considered for this purpose and are shown in Figures 6.17(A) and (B). They have the same enclosing rectangle, represented by a continuous line.

Their respective saturations lead to the values (A): $\frac{81}{140} \simeq 57.85\%$ and (B): $\frac{63}{140} = 45\%$, which highlights their respective density. However, saturation does not include shape characteristics. A first refinement is given by the histogram of the component. The histogram of a component is composed of the number of

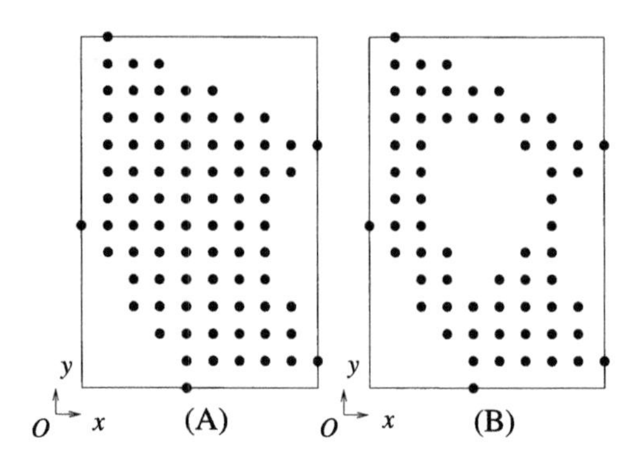

Figure 6.17 Saturation of a component

foreground pixels contained in each column (i.e. x-histogram) and line (i.e. y-histogram) of the component. Figure 6.18 illustrates the x- and y-histograms of the component shown in Figure 6.17(A) (the orientation and origin are illustrated in this figure).

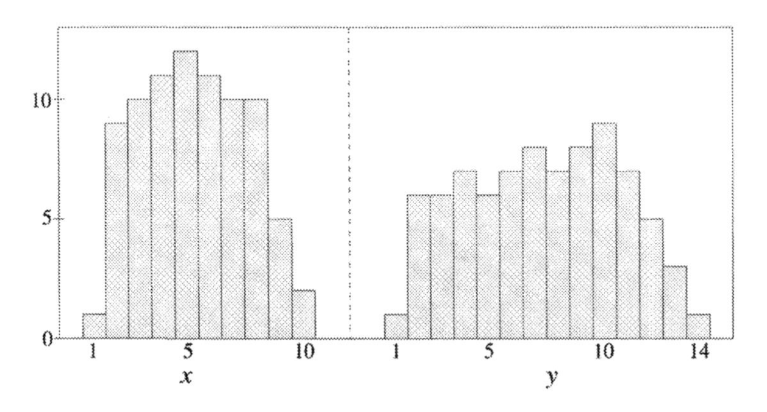

Figure 6.18 x- and y-histograms of the component seen in Figure 6.17(A)

Similarly, Figure 6.19 shows the x- and y-histograms of the component shown in Figure 6.17(B). By contrast, the location of the hole in the component can be characterised by a non-monotonic histogram (i.e. different modes in the histogram).

More precise characterisation of the shape of the component can be calculated. For example, based on the moments of the polygons, using continuous definitions, one can readily characterise best-fit rectangles and ellipses for a given component [62].

The compactness of a component without hole can be characterised by a single factor given by

$$\frac{(\text{Perimeter})^2}{4\pi(\text{Area})}$$

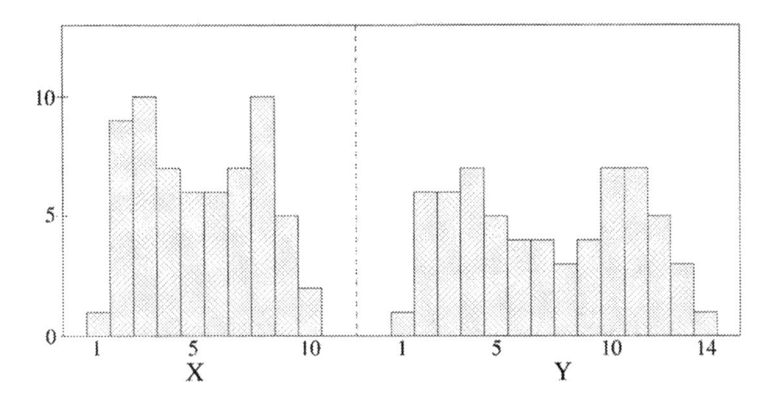

Figure 6.19 x- and y-histograms of the component shown in Figure 6.17(B)

Clearly, for a continuous disc, this value equals 1. Using the discrete circle presented in Figure 2.32 in Chapter 2, this factor equals $\frac{\left(\frac{72}{3}\right)^2}{4.\pi.46} \simeq 0.996$ with $a = 3$ and $b = 4$ as move lengths for calculating the perimeter (the value of the perimeter has then to be divided by a for consistency). By contrast, this factor equals $\frac{\left(\frac{116}{3}\right)^2}{4.\pi.63} \simeq 1.888$ for the component shown in Figure 6.17(A). The difference of these values thus highlights the shape difference between these two components.

Finally, the convexity of a component F may be measured by a factor involving a description of the component and its convex hull. For example, if $\langle F \rangle$ is the discrete convex hull of F and $\mathcal{A}_F$ (respectively $\mathcal{A}_{\langle F \rangle}$) is the area of F (respectively $\langle F \rangle$), the ratios $\frac{\mathcal{A}_F}{\mathcal{A}_{\langle F \rangle}}$ and $\frac{|F|}{|\langle F \rangle|}$ equal 1 if and only if $F = \langle F \rangle$ (i.e. is F is discrete convex) and therefore provide a measure for discrete convexity.

Such factors allow for a characterisation of properties related to the shape of the connected component in question. They can form a basis for the definition of more elaborate factors which are dedicated to a particular application. They may also be used when smoothing or segmenting a connected component as well as during a recognition process.

6.3.3.3 Curvature along the contour

Discrete curvature can be used for characterising circular objects. In Section 2.4, it was pointed out that digital straight segments and discrete circles could be characterised by constant curvature. Similarly, discrete curvature can be used to characterise angular pixels in the contour of a connected component. Figure 6.20 illustrates such a characterisation applied to the figure presented in Figure 6.13. The order-3 discrete curvature ρ_3 is plotted, starting from pixel p in the orientation shown in Figure 6.20. Clearly, angular pixels p_i in the contour are characterised by peaks in the curvature graph. Following the conclusions in

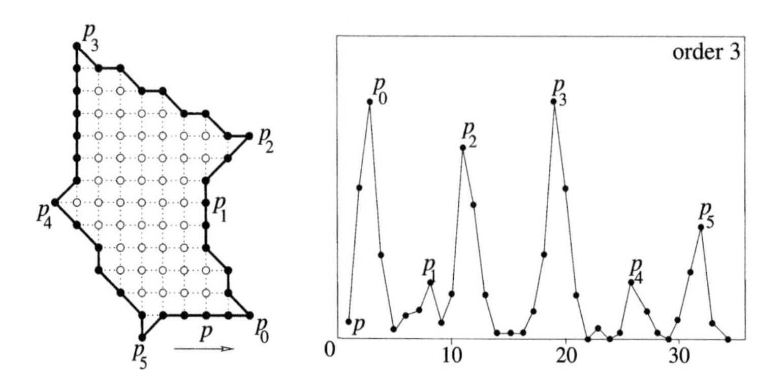

Figure 6.20 Curvature evolution along a contour

Section 2.4.3, the order of the curvature has to be adapted to the size of the component for such a characterisation. As result, a connected component may be represented by a continuous polygon based on angular pixels detected for further morphological characterisation.

Chapter 7

IMAGE THINNING

The interpretation of the content of an image relies on the definition of accurate representations on which an analysis process can be applied. Typically, for tackling problems such as image classification and image registration, invariants have to be designed. In Chapter 6, we presented some basic factors which can be associated with image components. However, these factors, descriptive as they may be, are often not specific enough for uniquely identifying the content of an image under study. In this chapter, we concentrate on the design of a class of models which will allow for such a characterisation. These models aim to map the image components onto their skeletons, corresponding to a well-established concept in image analysis. From their skeletons, reliable invariants can be designed.

This chapter is organised as follows. Firstly, Section 7.1 introduces the concept of skeleton, both informally and in a mathematical context. The most common mathematical skeleton models which have been proposed in the literature are reviewed. These models are introduced in continuous space in order to give formal justifications to skeleton properties. Based on these continuous models, Section 7.2 studies the discrete implementations of thinning processes which lead to component skeletons. Examples are given throughout this chapter to highlight the capabilities and shortcomings of the different approaches. Finally, Section 7.3 specialises on the class of binary *line* images. The relevance of image thinning is shown for this particular class of images and the processing is adapted to its properties. This section concludes with presenting the most well-known vectorisation procedures which complete the raster-to-vector problem.

7.1 Skeleton models

In this section, we introduce the concept of skeleton of an image. Processes which reduce the content of an image to its skeleton are generally referred to as image skeletonisation or image thinning processes. Image skeletonisation represents one of the most-used techniques for associating an image with an equivalent representation, in order to facilitate further analysis. It has generated an abundant literature and a comprehensive review of early techniques is given in [82]. Mostly with the aim of operating analysis techniques, image skeletons have been characterised by strict mathematical models. In Section 7.1.1, we first introduce the general principles of thinning operations. Section 7.1.2 then presents in detail mathematical skeleton models in the continuous space.

7.1.1 Image skeletons

We introduce thinning techniques in the context of image classification (or image registration) and image compression. Image classification and image registration imply the association of an image with a representation which is invariant against a given set of transformations. In the case of image classification, representatives associated with a set of images are then clustered into classes for further indexing. Image registration uses the representative of an image for finding the best match within a set of image models (e.g. for optical character recognition, OCR). In both cases, it essential that the image representatives are invariant against transformations induced, for example, by the acquisition process (see Chapter 4). Examples of such transformations include translation, scaling and rotation.

Image compression uses the image representative for storing the original image. It is therefore assumed that all the information needed for recovering the original image is contained within its representative only. In other words, one should be able to define a "reconstruction" process for uniquely mapping the image representative onto the original image. At this stage, a loss of information may be considered for simplifying further the form of storage. By contrast with the image compression techniques presented in Section 4.2.2, such techniques take advantage of global image properties to accurately reduce the amount information stored (i.e. *adaptive* techniques rather than *blind* techniques).

Amongst a diversity of image representatives (e.g. component borders), image skeletons are found to be very efficient in terms of representation and description of the original image. The underlying idea behind image skeletonisation is to define a thin central structure which uniquely represents components in the image. Intuitively, the information contained in a ribbon-like component is mostly contained within such a structure. See Figure 7.1 for example, where the skeleton of each component is represented by a thick continuous line.

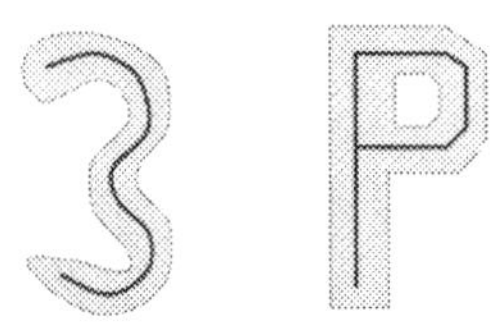

Figure 7.1 Image skeleton

This is generalised to generic components via the definition of thinning processes. More formally, the following description of a perfect image skeleton is as follows.

Proposition 7.1: Skeleton

Let F be the foreground of a binary image. If S is the skeleton of F, then:

(i) S is at a central position within F. In particular, S is totally contained in F.

(ii) S is one-pixel wide.

(iii) S has the same number of connected components as F.

(iv) S^{c} has the same number of connected components as F^{c}.

(v) S allows for the reconstruction of F.

Remark 7.2:

Conditions (iii) and (iv) are generally summarised by saying that S and F are homotopic. In other words, the topological (upper-level) structures of S and F are equivalent (i.e. can be mapped onto one another by a continuous transformation).

It may not be possible to fulfil all the above conditions at the same time. Therefore, depending on the application in question, emphasis will be placed on particular conditions while relaxing the other conditions. For example, in the case of compression, condition (v) proves fundamental whereas condition (ii) can be ignored. By contrast, for image classification or registration, condition (v) can be relaxed and emphasis is mostly placed on the remaining conditions.

The next section reviews well-known mathematical formulations for the above conditions. For the sake of simplicity, these formulations are first set in the continuous space. It will be our aim in Section 7.2 to map continuous skeleton models into discrete space and to propose algorithms for characterising their solutions.

7.1.2 Mathematical models

Mathematical models for image skeletons are based on the study of a continuous set F. The aim is to characterise a continuous set S which represents the skeleton of F. Without any loss of generality, we can assume that F contains one connected component only. However, F^{c} may contain more than one connected component (i.e. F may contain holes). Under this assumption, S has clearly to be connected in order to fulfil conditions (iii) and (iv) in Proposition 7.1.

Different models have been proposed to characterise S. In this section, we review three models whose discrete implementation will make use of the discrete methods described in the previous chapters. For the sake of completeness, we also review a model based on Voronoi diagrams which has been proposed for obtaining continuous skeletons.

7.1.2.1 *Blum's model*

An early model for characterising the skeleton of a continuous component F was proposed by Blum in [8, 10]. To simplify the formulation, the component F is assumed to have a smooth border (i.e. a tangent vector can be defined at each point of its border). F is also assumed to contain no holes. However, as we will see, this model can readily be extended to any component F.

Blum's model formulates the construction of F in terms of its generators, namely a curve S and a structural element. In the real plane $\mathbb{R}^2$, this model is given as follows.

Definition 7.3: Blum's model of a component

Given a curve S defined by $S : [0,1] \longrightarrow \mathbb{R}^2$ and a radius function $r : [0,1] \longrightarrow \mathbb{R}^+$, F is defined from S and r by

$$F(S,r) = \left\{ \alpha \in \mathbb{R}^2 \text{ such that } \exists s \in [0,1] \text{ such that } d_{\mathrm{E}}(S(s),\alpha) \leq r(s) \right\}$$

By definition, the structural element is a Euclidean disc $\Delta_{\mathrm{E}}(\rho)$ whose radius is given by $\rho = r(s)$ when centred at each point $S(s)$ of the curve. In this respect, the components F given by this model are referred to as "Blum's ribbon". Hence, an equivalent characterisation of F is given by Proposition 7.4.

Proposition 7.4: Blum's ribbon

Given a curve S defined by $S : [0,1] \longrightarrow \mathbb{R}^2$ and a radius function $r : [0,1] \longrightarrow \mathbb{R}^+$, F is defined from S and r by

$$F(S,r) = \bigcup_{s \in [0,1]} \Delta_{\mathrm{E}}(S(s), r(s))$$

where $\Delta_{\mathrm{E}}(S(s), r(s))$ is the Euclidean disc of radius $r(s) \in \mathbb{R}^+$, centred at $S(s) \in \mathbb{R}^2$.

We illustrate this model by Figure 7.2 where the component is equivalently represented by the thick curve. Instances of structural elements (dashed discs) are shown at different locations on this skeleton.

We now give a fundamental definition which will help us in characterising the skeleton of a continuous set F using the above model.

Definition 7.5: Maximal disc

A Euclidean disc $\Delta_{\mathrm{E}}^{\max}(p,r)$ of radius $r \in \mathbb{R}^+$ centred at $\alpha \in F$ is said to be maximal in F if and only if

(i) $\Delta_{\mathrm{E}}^{\max}(\alpha,r)$ is totally contained in F.

(ii) There is no other disc $\Delta_{\mathrm{E}}(\beta, r,)$ totally contained in F which contains $\Delta_{\mathrm{E}}^{\max}(\alpha,r)$.

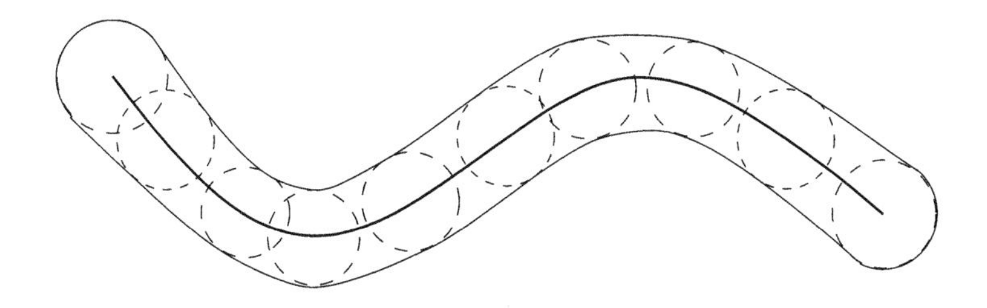

Figure 7.2 Blum's model of a ribbon-like image

More formally, $\Delta_E^{max}(\alpha, r)$ *is a maximal disc in* F *if and only if*

(i) $\Delta_E^{max}(\alpha, r) \subseteq F$.

(ii) For any $\beta \in F$ *and for any* $r' \in \mathbb{R}^+$ *such that* $\Delta_E(\beta, r') \subseteq F$, $\Delta_E^{max}(\alpha, r) \not\subseteq \Delta_E(\beta, r')$.

An equivalent characterisation of a maximal disc in F is as follows.

Proposition 7.6: Maximal disc

A disc $\Delta_E^{max}(\alpha, r)$ *is maximal in* F *if and only if it is a tangent with the border of* F *in at least two points. In this respect,* $\Delta_E^{max}(\alpha, r)$ *is often called a "bitangent disc" in* F.

By definition, each disc $\Delta_E(S(s), r(s))$ described in Definition 7.3 is maximal in F. Moreover, the set of discs $\Delta_E(S(s), r(s))$ contains all maximal discs in F. In other words, $\alpha \in \mathbb{R}^2$ is a centre of a maximal disc in F if and only if there exists $s \in [0, 1]$ such that $S(s) = \alpha$. Hence, the characterisation of a set F constructed according to Blum's model given in Definition 7.3 is as follows.

Proposition 7.7: Blum's skeleton

Given a set F*, the generators of* F *are the set* $S(s)$ *of centres of maximal discs* $\Delta_E^{max}(S(s))$ *and their corresponding radius function* $r(.)$.

The mapping from F to $S(.)$ and $r(.)$ is referred to as the Symmetric Axis Transform (SAT) in [8] and Medial Axis Transform (MAT) in most of the subsequent literature (e.g. [100]).

7.1.2.2 Wave propagation

Montanari [98] proposed an alternative characterisation and method for obtaining the skeleton of continuous set F. The idea is to propagate inside F wavefronts at constant speed from each point of the border of F in a direction locally perpendicular to the border of F. At some instant t, two or more wavefronts will collide in a point which is then said to belong to the skeleton S of

F. The value of t at which each skeleton point is created is then stored for reconstruction via a inverse propagation.

By simple geometric considerations, it is easy to show that this model is exactly equivalent to the previous model. Each skeleton point α is equidistant to at least two points β_i of the border of F, creating directions $[\alpha, \beta_i]$ perpendicular to the border of F at each point β_i, thus characterising $d_{\mathrm{E}}(\alpha, \beta_i)$ as the radius of a maximal disc in F.

Blum [9] called this transformation the Grassfire Transform by analogy with a fire set at each point of the border of F and propagating inside F. In this context, skeleton points correspond to quench points.

Figure 7.3 illustrates how this model works conceptually. The dashed lines symbolise the wavefronts and the thick line symbolises the resulting skeleton. The arrows show that waves are propagated along a direction which is locally perpendicular to the border.

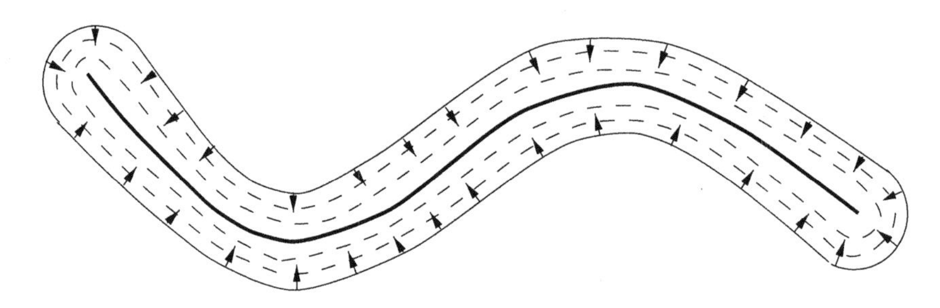

Figure 7.3 Wavefront propagation model

7.1.2.3 Local width

In [151], a different geometrical approach is used to characterise skeleton points. The idea behind this model is to calculate an approximation of the characterisation given by the previous models. The contour of the pattern is assumed to be smoothed and to form a non-self-intersecting closed loop. Then, a trapezoid, $((\alpha_1, \alpha_2, \beta_2, \beta_1)$, say – see Figure 7.4), is described whose two opposite sides (e.g. $[\alpha_1, \alpha_2]$ and $[\beta_1, \beta_2]$) each belong to the opposite sides of the contour. One remaining side of the trapezoid (e.g. $[\alpha_1, \beta_1]$) is called a base segment. The shortest segment between $[\alpha_2, \beta_2]$ (opposite to $[\alpha_1, \beta_1]$) and $[\alpha_1, \beta_2]$ and $[\alpha_2, \beta_1]$ (diagonal lines of the trapezoid) is called a minimum-base segment. Under certain conditions (e.g. the width of $(\alpha_1, \alpha_2, \beta_2, \beta_1))$ and geometric approximations, the midpoint γ of the minimum-base segment is shown to be part of the skeleton.

Clearly, in this model, the smaller $|d_{\mathrm{E}}(\alpha_1, \beta_1) - d_{\mathrm{E}}(\alpha_2, \beta_2)|$, the closer γ is to the exact skeleton. The approximation made in this model relies on the fact that this characterisation is to be applied in the discrete space (i.e. where F is a set of connected pixels). In this case, under certain criteria, we shall see in Section 7.2 that this technique produces valid skeletons.

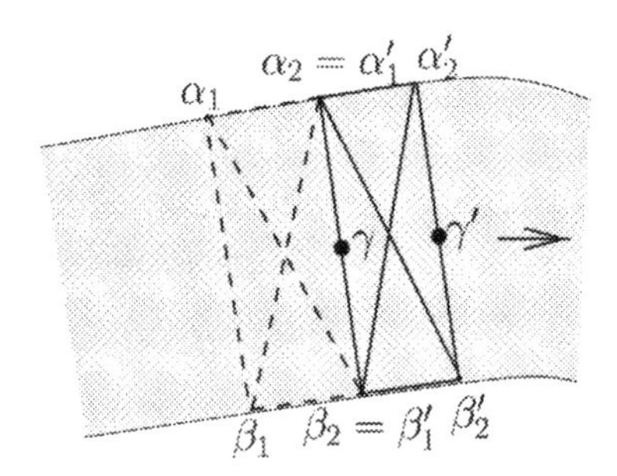

Figure 7.4 Minimum-base segment algorithm. •: Skeleton point

7.1.2.4 Skeletons based on Voronoi diagrams

By definition, the skeleton of a component F is the set of points in F which are equidistant from at least two points of the border of F. In this respect, skeletons can be characterised as part of the border of cells in the Voronoi diagram associated with the border of F. However, this definition generally involves the construction of generalised Voronoi diagrams where seeds are not points but rather curves. This approach therefore has no practical implementations in the continuous space. However, we will see that, in the discrete space, the close relationship between distance maps and Voronoi diagrams will make this approach implicitly equivalent to the above.

7.2 Thinning algorithms

Owing to the discrete nature of binary images, skeleton models have to be adapted in a discrete context. This is done by using tools and properties described in the previous chapters. Discrete distances will help in measurements, whereas connectivity will prove fundamental for developing and evaluating these algorithms. The aim is to characterise a skeleton as a set of pixels S within the foreground of the original image. In this section, discrete equivalents to the models presented are found and described.

These techniques can be divided into two main categories. The first class of techniques, referred to as skeletonisation techniques, first characterise a set of "skeletal" pixels eligible to belong to the final skeleton. Further processing based on the conditions given in Proposition 7.1 is needed to obtain the skeleton set S.

The algorithms of the second class are called thinning algorithms and process the image iteratively until the skeleton set S is obtained, while maintaining the validity of conditions in Proposition 7.1 throughout the process.

In Section 7.2.1, we present a classical algorithm which relies heavily on discrete distances and distance maps. This algorithm belongs to the class of skeletonisation algorithms since it first characterises a skeletal set which is then post-processed. Section 7.2.2 then proposes to use the well-structured context of mathematical morphology to perform thinning. The most commonly used operators are defined and their implementation discussed. As a complement, a different approach which is based on the minimum-base segment algorithm (see Section 7.1.2) is presented in Section 7.2.3.

7.2.1 Centres of maximal discs

This approach relies on Blum's model of a binary component F (see Definition 7.3). It makes direct use of the result presented in Proposition 7.7. The idea behind this technique is to characterise a set of skeletal pixels (i.e. eligible for being skeleton pixels) as the set of centres of maximal discs in the component. The skeletal set is then processed in order to fit the conditions given by Proposition 7.1. In particular, connectivity and one-pixel width are ensured via deletion or addition of pixels in this set.

The use of a specific discrete distance d_{D} has first to be decided. It is commonly admitted that chamfer distances allow for an efficient characterisation of the centres of maximal discs. In Proposition 5.5(i), it was noted that the value of the distance transformation of F at p, $\mathrm{DT_D}(p)$, indicates the radius of the largest discrete disc centred at p and totally contained in F. Combining this result with Proposition 7.7, centres of maximal discs in F can clearly be characterised as pixels in F which correspond to local maxima in the distance map of F.

Remark 7.8:

Strictly speaking, centres of maximal discs are local maxima of the distance map if and only if the basic move length (i.e. a) is 1. When extending discrete distances, this property is lost but the name is kept.

When using discrete distances, Proposition 5.7 describes the propagation of local distances with the distance mask. In short, a pixel $q \in F$ participates in the propagation of the distance transformation value from a border pixel r to a pixel p (neighbour of q) if and only if $\mathrm{DT_D}(p) = \mathrm{DT_D}(q) + d_{\mathrm{D}}(p,q)$, where $\mathrm{DT_D}(q) = d_{\mathrm{D}}(q,r)$. A pixel is a centre of maximal disc if it does not participate in the propagation of the distance transformation value at any pixel. In summary, the following proposition holds.

Proposition 7.9: Centre of a maximal disc [2]

Given a set F of pixels and a discrete distance function $d_{\mathrm{D}}(.,.)$ based on a neighbourhood $N_{\mathrm{D}}(.)$ defined on F, a pixel $p \in F$ is a centre of maximal disc in F

with respect to d_D, *if and only if*

$$\mathrm{DT_D}(q) < \mathrm{DT_D}(p) + d_D(p,q) \quad \forall q \in N_D(p)$$

The radius of the maximal disc is then given by $\mathrm{DT_D}(p)$.

The set of centres of maximal discs forming the skeletal set is therefore characterised using a local search within the distance map in one complete scan of the image. Because of this local property, conditions *(i)* to *(iv)* of Proposition 7.1 may not be satisfied. However, condition *(v)* of Proposition 7.1 is satisfied. In other words, complete reconstruction of the original image is ensured.

Remark 7.10:

By definition, a centre of maximal disc is a pixel located on the ridge of the scalar field created by the distance map. At a centre of maximal disc, border influence therefore switches from one side to the opposite. By duality, considering borders as generators of a generalised Voronoi diagram, the set of centres of maximal discs therefore constitute the borders of Voronoi cells. Figure 7.5 shows an example of a binary image and its associated distance map.

In this representation, each pixel is connected to its closest border pixel. White-like areas corresponding to ridges in the distance map show locations of centres of maximal discs and constitute edges of the Voronoi diagram. By definition they also form the skeleton of this binary component.

Figure 7.5 Discrete Voronoi diagram deduced from the distance map

Depending on the application considered (e.g. compression or representation), the skeletal set may be post-processed in different ways to emphasise different conditions in Proposition 7.1.

If thinning is performed with the aim of representation, conditions *(ii)* to *(iv)* need to be followed as closely as possible. Some pixels are therefore to be removed or added in the skeletal set to form the final skeleton. Such post-processing mostly relies on the inspection of the neighbourhood of each pixel in the skeletal set. The use of connectivity and crossing numbers as defined in Definition 6.6 then helps in deciding whether to add (or remove) a specific pixel to (or from) the skeletal pixel. More generally, the definition of neighbourhood masks (i.e. templates) which show allowed and forbidden patterns in the skeletal set allows for respecting specific constraints. This is formally described, next, using a morphological approach.

As further processing, a pruning process will remove small branches from the skeletal set, considered as spurious. In this respect, such step is often called a "beautifying step".

When the aim of thinning is compact storage (i.e. compression), redundancy has to be removed from the skeletal set. The skeletal set as it is given in Proposition 7.9 allows for exact (complete) reconstruction of the original image. However, a subset may be defined such that it still allows for exact reconstruction while including fewer pixels (e.g. see [52, 77, 87]).

7.2.2 Morphological approach

When using a morphological approach, thinning is related to the wave propagation model proposed earlier. In the discrete space (i.e. using pixels), wavefronts are propagated using operators defined in Definitions 6.2 and 6.4. In this section, we simply introduce the concepts related to morphological skeletons and morphological thinning. For a thorough study of morphological thinning and skeletons, the reader is referred to [48, 126, 149, 150].

Definition 7.11: Morphological skeleton

Let F be the foreground of a binary image. The morphological skeleton S of F is defined by

$$S = \bigcup_{n=0}^{\infty} S_n$$

where

$$S_n = F_n \setminus \text{opening}(F_n) \quad \textit{and} \quad F_n = \text{erosion}^n(F)$$

The notation $\text{erosion}^n(F)$ *denotes n successive applications of the* erosion(.) *operator to F:*

$$\text{erosion}^n(F) = \overbrace{\text{erosion}(\text{erosion}(\ldots\text{erosion}(}^{n \text{ times}} F)\ldots))$$

Remark 7.12:

By definition of F_n, the union is clearly reduced to

$$S = \bigcup_{n=0}^{N} S_n$$

where N is the smallest integer value such that $F_{N+1} = \emptyset$.

Example: Morphological skeleton

We present an example of morphological skeletonisation of a simple binary image (Figure 7.6). The original image is a square-like shape added with a small branch. Here, $N_D = N_4$.

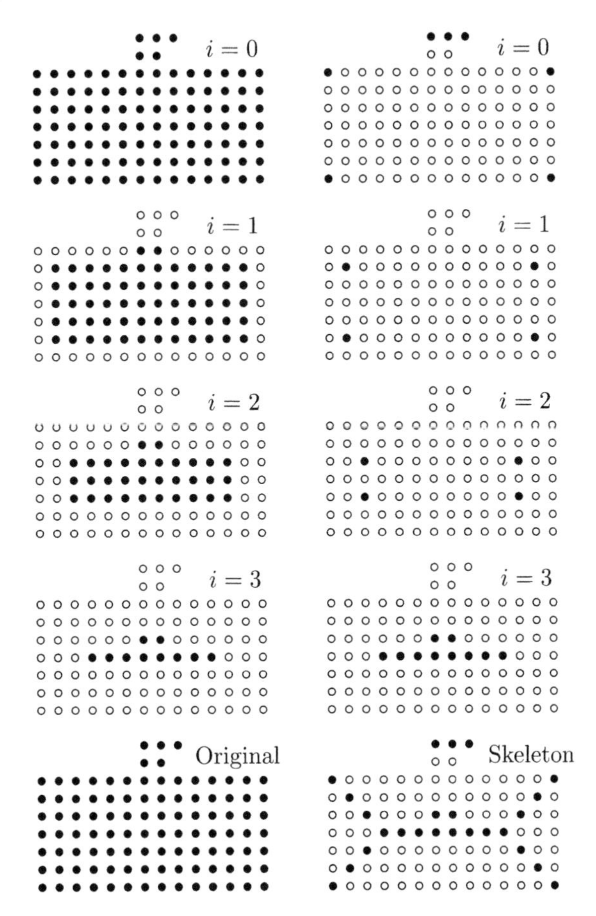

Figure 7.6 Morphological skeleton

The left column shows successive sets F_n and the right column shows successive sets S_n. In that case $N = 3$ since $F_4 = \emptyset$. The bottom row compares F and S. Note that, in this case, S is a disconnected set.

Use of the definition of a morphological skeleton in its digitised form may result in a disconnected skeleton set S even if the original component F is connected. Therefore, the definition of a morphological thinning operator has been introduced.

The thinning operator defined in the context of mathematical morphology relies on the definition of the "hit or miss" operator denoted by $\circledast$

Definition 7.13: Hit or miss operator ($\circledast$)

Consider F the foreground of a binary image and B a particular configuration of the neighbourhood around a point. By extension of the context defined in Definitions 6.2 and 6.4 and Remark 6.3, B is referred to as a structuring element. $F \circledast B$ is formed by pixels in F whose neighbourhoods match the structuring element B.

Example: Hit or miss operator

Let B be the structuring element shown in Figure 7.7(A). The arrow points to the pixel on which the structuring element will be centred (i.e. the origin of the structuring element), and • and ○ denote foreground and background pixels, respectively. Pixels which do not appear are not taken into consideration. Let F be as given in Figure 7.7(B). Then, $F \circledast B$ is shown in Figure 7.7(C). Figure 7.7(D) highlights the fact that black pixels in $F \circledast B$ correspond to pixels in F whose neighbourhood configurations match B.

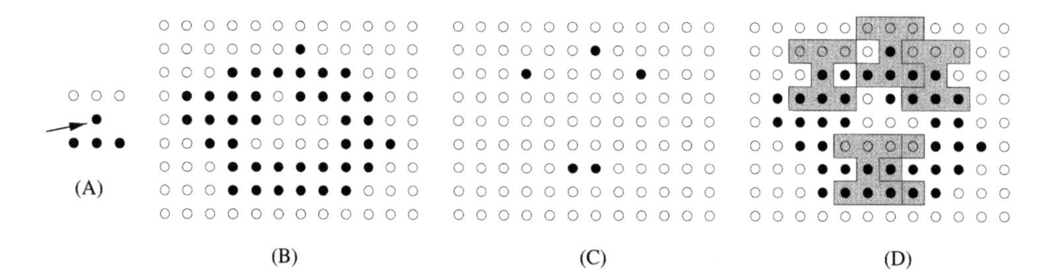

Figure 7.7 Hit or miss operator. (A) B, the arrow points to the origin pixel. (B) F. (C) $F \circledast B$. (D) Locations where B is matched

In this context, morphological thinning can be defined with respect to a structuring element B.

Definition 7.14: Morphological thinning

Given a structuring element B and the foreground F of a binary image,

$$\text{thinning}_B(F) = F \setminus F \circledast B$$

In other words, $\text{thinning}_B(F)$ *is the set of pixels in F whose neighbourhoods do not match B.*

Using this operator, binary image thinning is typically performed by applying in parallel rotated versions of the structuring element B. A widely used structuring element is that shown in Figure 7.7(A). Figure 7.8 illustrates the difference between the result using thinning(.) operator (B) (B is as shown in Figure 7.7(A)) and the morphological skeleton (C) of a binary image (A).

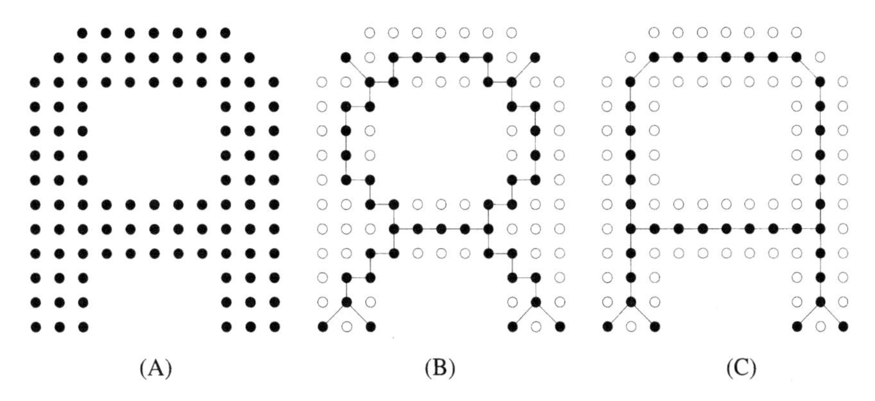

Figure 7.8 Morphological thinning versus skeleton. (A) F. (B) thinning$_B(F)$. (C) Morphological skeleton (Definition 7.11)

As a complement to these operators, the context of mathematical morphology provides pruning operators to remove spurious branches from the final thinned set. This is particularly useful when thinning noisy images so that reconstruction will be more symmetric and smoothed. Pruning is typically based on the hit or miss operator to exploit conditions within the neighbourhood. This clearly corresponds to using connectivity and crossing numbers and is equivalent to the beautifying step mentioned earlier.

7.2.3 Minimum-Base Segment algorithm

In the discrete space, skeletonisation based on local width simply uses pixels as border points and follows the borders of the component in directions allowed by the connectivity relationship given in the current context. The algorithm detailed in [151] shows some shortcomings which are overcome in [91].

The main advantage of such a discrete approach is that it creates pairs of border pixels to form local width lines. These lines have been shown to be optimal for local width measurements and therefore for analysing the image. An example is illustrated in Figure 7.9 where the skeleton and dual width lines are shown for a ribbon-like image.

7.3 Binary line images

Binary images are typically used in applications where the shapes of the foreground components are meaningful. This is the case, for example, in images

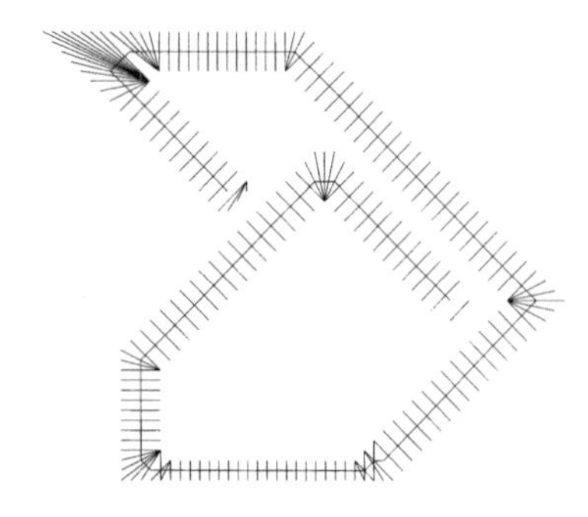

Figure 7.9 Skeleton and local width measurement

representing an alphabet (e.g. characters and sign language) or a shape allowing identification (e.g. fingerprint). What all these images have in common is that their skeletons are almost as meaningful as the images themselves. By contrast, it is not clear how to identify a blob-like shape from its skeleton without reconstructing it. This is emphasised schematically in Figure 7.10, where images of different types and their respective skeletons are presented.

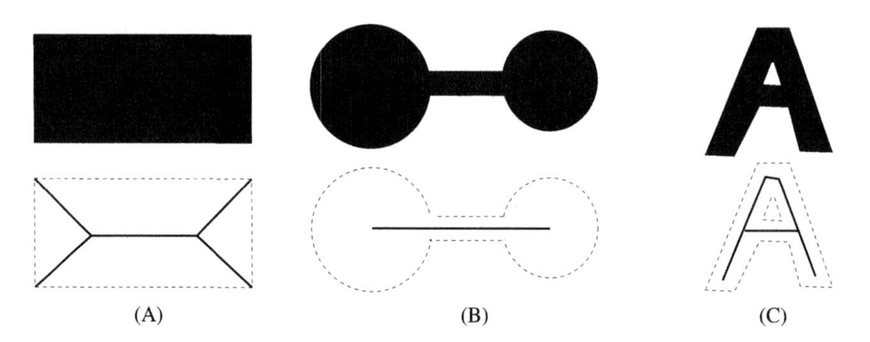

Figure 7.10 Examples of images and their skeletons

Figure 7.10(C) shows a typical example of what we call a binary *line* image. It is a binary image composed of strokes where the width information is irrelevant. The representation of such an image can accurately be done by finding its skeleton. From this feature, the image can be interpreted without the need for reconstruction. It is actually proven that such a representation is more reliable than the image itself (e.g. for OCR applications). In this section, we study the class of binary line images via the significance of their skeletons (Section 7.3.1).

Remark 7.15:

In this study, we leave aside the case of half-tone pictures, which are a particular class of binary images which tends to belong to the class of gray scale images.

To complement the approaches developed in the general case (Section 7.2), we then introduce a new graph-theoretic approach for skeletonisation and show that it offers an alternative efficient route for the analysis of binary line images (Section 7.3.2).

7.3.1 Significance of skeletons

Examples of binary line images are characters (and more generally symbols of most alphabets), road maps, engineering drawings and fingerprints (see Figure 7.11).

Figure 7.11 Instances of binary line images

All these images have in common the fact that they represent a network composed of strokes. They are analysed via the reduction of these strokes to unit width, therefore forming their upper-level graph structure. From then on, a description of the nodes (possibly node locations) and their inter-relationships (upper-level connectivity) can easily be derived.

It is therefore clear that, in this context, the skeleton as defined in Proposition 7.1 is of fundamental importance for binary line image understanding. In Section 7.2, it is argued that, depending on the application considered, some of the conditions in Proposition 7.1 can be relaxed. In the case of binary line image skeletonisation, the homotopicity of the skeleton with the original image is of crucial importance (i.e. emphasis will be placed on conditions (iii) and (iv)). To highlight this, consider the following example:

Example: Binary line image skeletonisation

Say that a binary image has been extracted from a satellite image that contains roads only. When thinning this binary image, the road network can be recovered from the upper-level graph structure of the binary line image. From then on, it is straightforward to conceive of an application which allows a user to find the shortest route from one place to another. This implies that length measurements take place within the strokes and that their connectivity needs to be ensured in order for this application to be usable.

Stroke length measurement calls for following each skeleton branch in some way. From techniques defined in Section 7.2, the skeleton is given as a raw set of pixels. No structure is provided to investigate the skeleton. In the next section, we introduce a graph-theoretic approach, which we show is particularly well-suited for the type of application described in the above example.

7.3.2 Graph-theoretic approach

In the particular case of binary images, connectivity plays an important role in thinning operators. In Chapter 3, a graph-theoretic context for binary image processing has been introduced. The aim of this section is to extend this context and show that graph theory may be a powerful complement to classical approaches.

7.3.2.1 Principle

A skeleton, as given in Proposition 7.1 is a thin subset located at a central position within the foreground of the original image. Graph-theoretic thinning will consist of two major steps. Firstly, centrality within the foreground components is to be characterised. This is done using distance maps detailed in Chapter 5. In this respect, it will be shown that the skeleton obtained from this characterisation follows the line of that obtained using Blum's model. In Chapter 3, it was shown that connectivity could naturally be mapped onto a low-level graph structure via the grid graph. This structure will be exploited for preserving connectivity during thinning.

In Chapter 5, distance maps were defined as the set of values of a distance transform at every pixel of the image. Following Blum's model, this value indicates the radius of the largest disc completely contained in the foreground and centred at that point and therefore indicates how central a point is within the foreground component. Graph-theoretic thinning uses this centrality measure to determine a central subset of pixels which will form the skeleton as follows.

Without any loss of generality, we consider an image whose foreground is composed of one connected component only. Let $G = (V, A)$ be the grid graph associated with that component (i.e. V is the set of foreground pixels) with respect to a given connectivity relationship. To each pixel p in V, we associate $\mathrm{DT_D}(p)$, the value of the value of the distance transform associated with the discrete distance D at that pixel. To each arc $(p, q) \in A$ joining two pixels p and q, we associate a weight $w(p, q)$. The value $w(p, q)$ will be calculated as a function of the centrality measure at p and q. A simple example of such weight is

$$w(p, q) = \frac{\mathrm{DT_D}(p) + \mathrm{DT_D}(q)}{2} \quad \forall p, q \text{ such that } (p, q) \in A$$

Let $P = \{p_1, \ldots, p_n\}$ be a path in G; the total weight of this path is therefore

$$w(P) = \sum_{i=1}^{n-1} w(p_i, p_{i+1}) = \frac{\mathrm{DT_D}(p_1) + \mathrm{DT_D}(p_n)}{2} + \sum_{i=2}^{n-1} \mathrm{DT_D}(p_i)$$

In other words, the weight of a path P is the sum of the values of the centrality measures at pixels corresponding to the vertices that compose P. Therefore, it is natural to characterise a central path (i.e. the skeleton) as a maximum weighted path in G. This maximum weighted path is typically found by using a shortest path procedure with the inverted weight $w_{rmI}(p, q) = \max_{(u,v)\in A} w(u, v) - w(p, q)$ as arc length.

Following the same principle, a more sophisticated weighting scheme can be designed. It may be necessary to include arc lengths in this weighting scheme as

$$w(p, q) = \frac{\mathrm{DT_D}(p) + \mathrm{DT_D}(q)}{2} l(p, q)$$

where $l(p, q) = d_\mathrm{D}(p, q)$ is the length of the arc (p, q) with respect to the discrete distance d_D chosen. This is, for example, the case when using a 5×5 neighbourhood and $d_{a,b,c}$ to compensate for the fact that c-moves are skipping some grid lines.

By definition, a path is a connected set of pixels having a unit width. Therefore, conditions (i) to (iv) in Proposition 7.1 will be automatically satisfied by the definition of our graph-theoretic skeleton. In this context, the problem of thinning an image therefore reduces to finding the end point of strokes that compose the image. The next section outlines a possible technique for finding such pixels and for splitting the foreground into non-branching parts.

Assuming that F is a non-branching component of the foreground of a binary image, the two end points of the skeleton of F are easily characterised as extremities of the longest shortest path in F. In practice, they are found by calling a shortest path procedure twice. Starting from a random vertex r, the shortest path spanning tree rooted at that vertex is constructed and the farthest vertex u_a is selected (i.e. u_a is such that $P(r, u_a)$ is maximum in the shortest path spanning tree). The shortest path procedure is called from u_a again to characterise u_b in the same manner. The final skeleton is finally found by finding the minimum weighted path from u_a to u_b. This procedure is summarised in Figure 7.12.

The procedure described above assumes the skeleton to be non-branching. There is therefore a need for a preprocessing step which decomposes the original binary line image into ribbon-like components. Such a technique based on graph theory is presented in the next section.

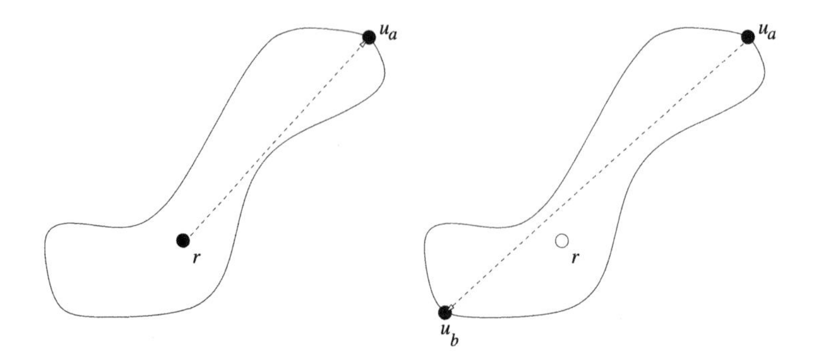

Figure 7.12 Extremities of a ribbon-like image

Remark 7.16:

Note that the procedure described above can readily be used as a beautifying step in classical methods. The preprocessing step then consists of using classical thinning techniques and the graph-theoretic approach ensures cleaning and analysis of the final skeleton.

7.3.2.2 Upper-level structure of a binary line image

A binary line image has been defined as a union of ribbon-like parts. The underlying network structure represented by such an image is called the upper-level structure of the image. The method outlined here takes advantage of graph-theoretic algorithms for retrieving this structure and therefore characterising a large part of the information contained in the image. Figure 7.13 illustrates this methodology.

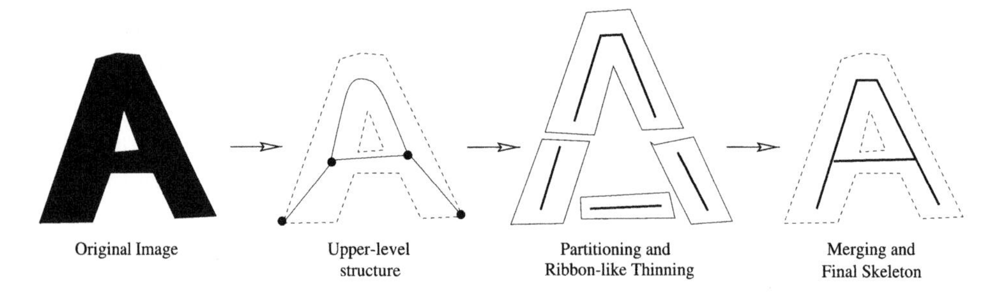

Figure 7.13 Steps of graph-theoretic thinning

In a generic binary line image F (i.e. possibly branching) two points are first characterised as the extremities of the longest shortest path as detailed earlier. Using a k-shortest path procedure, all possible routes between these two points can then be computed. The upper-level structure will be the union of all these routes. From then on, skeleton branches can be characterised using a given weighting scheme and the technique presented in the previous section.

We conclude this part with detailing a practical example of graph-theoretic thinning applied to the image of a transportation network.

Example: Graph-theoretic thinning

Consider the image F shown in Figure 7.14, which sketches a transportation network.

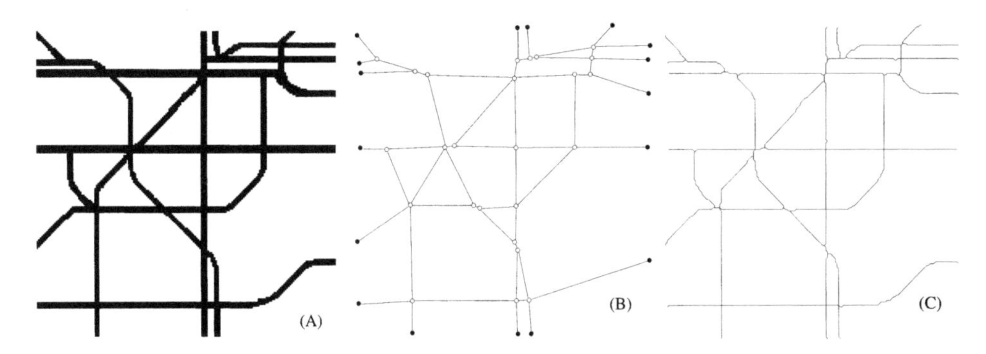

Figure 7.14 Graph-theoretic thinning of a network image

Using a k-shortest path technique, the upper-level of F is first retrieved (Figure 7.14(B)) so that thinning can be achieved. Figure 7.14(C) shows the result of such a process. In this case, S is represented by the set of arcs joining pixels in S to emphasise its connectivity and its unit-width. This example shows how shortest path procedures can be used for avoiding spurious branches that may be caused, for example, by noisy borders.

From this example, it is clear that the analysis of F will equivalently be performed on S.

7.3.3 Vectorisation

A typical application of the processing of binary line images is the vectorisation of such images. For example, an engineering drawing composed of lines only can efficiently be represented by the set of extremities of these lines (referred to as major points) and their inter-connections. The aim now is therefore to characterise this information from the skeleton of the original image. Line extremities can be divided into two subsets, namely, junctions or end points and angle points. Junctions and end points are readily given from the thinning process described above. In this section, we will concentrate on describing methods which recover angle points in a line image. As output, a set of points in the two-dimensional space and the structure of the image are obtained. From then on, geometrical operations can be performed accurately on this information. Moreover, such an output allows for the combination of different images. Three-dimensional reconstruction from two-dimensional views (CAD) becomes possible. Another application may also aim at measuring the changes of feature parameters over a sequence of images.

The aim of this section is to present techniques to approximate chain-code sequences by polygonal lines. Characteristics describing an automated vectorisation system are given as follows.

- The location of the resulting major points are accurate within the image.
- Recovered features should give an accurate description of the input image.
- Possible noise should be taken into account during the processing.

Different approaches have been taken for solving this problem. They mostly reduce to the following.

Given a chain-code sequence $\{c_i\}_{i=1,\ldots,n}$ (see Definition 2.3) and assuming that the starting point of this sequence is given by the origin $p_0 = (0,0)$, $\{c_i\}_{i=1,\ldots,n}$ can uniquely be mapped onto the digital arc $P_{p_0p_n} = \{p_i\}_{i=0,\ldots,n}$ where a segment $[p_{i-1}, p_i]$ represents the chain code c_i. In other words, the given of the chain-code sequence $\{c_i\}_{i=1,\ldots,n}$ is equivalent to a given set of points $\{p_i\}_{i=0,\ldots,n}$. For approximating the polygonal curve defined by $P_{p_0p_n}$, a set of points $\{p'_i\}_{i=0,\ldots,m}$ is to be characterised under the following conditions.

Proposition 7.17: Characterisation of the polygonal approximation

(i) Conservation: $\{p'_i\}_{i=0,\ldots,m} \subset \{p_i\}_{i=0,\ldots,n}$ *and, if* $p'_i = p_j$ *then* $p'_{i+1} \in \{p_l\}_{l=j+1,\ldots,n}$.

(ii) Relaxation: *The distance between point* p_k *and the approximating segment* $[p'_i, p'_{i+1}] = [p_{k_1}, p_{k_2}]$ *such that* $k_1 \le k \le k_2$ *is below a certain threshold.*

(iii) Optimality: m *is minimum among all sets satisfying the above criteria.*

These conditions lead to two major problems. The first one is to select the measure that defines the relaxation and the second issue is to find an optimal value of the number m of approximating segments with respect to this measure.

7.3.3.1 "Parallel-strip" error criterion

The "parallel-strip" error criterion [163] gives a measure of how close each approximating segment $[p'_i, p'_{i+1}] = [p_{k_1}, p_{k_2}]$ is from $\{p_i\}_{i=0,\ldots,n}$. The idea is to measure the distance between every point p_k and the approximating segment $[p'_i, p'_{i+1}] = [p_{k_1}, p_{k_2}]$ such that $k_1 \le k \le k_2$.

In the work presented in [41,163] (see also [96]), optimality is defined as the minimal total deviation between the original and the approximating polygonal line. Before presenting a graph-theoretic approach to optimality as defined in Proposition 7.17, we present an alternative approach.

7.3.3.2 Recursive partitioning

A technique for partitioning a digital arc into piecewise linear curves is proposed in [86]. The digital arc is hypothesised to be represented by a straight line passing through its extremities. The chain is then recursively divided into two sub-parts at its point of highest curvature (see Section 2.4) which corresponds to the point of maximal deviation from the hypothetical segment. This process is stopped when such a sub-part reaches a given minimum length. The result is a binary tree in which nodes are sub-parts of the original digital arc created during the decomposition. This tree is then used as a representation of the original digital arc. This work has been extended in [144, 170] for fitting arcs or ellipses. Reference [143] gives a summary and a comparison of curve-fitting approaches for solving the polygonal approximation problem.

7.3.3.3 Graph-theoretic approach

The optimality of m, the number of approximating segments, may be resolved using a graph-theoretic approach, consistent with that used throughout this book. The outline of such a technique using graph theory for minimising the number of approximating segments is as follows.

A "visibility" relationship is first to be defined. It determines for each point p_i the point p_j for which any approximating segment $[p_i, p_j]$ satisfies the condition (ii) in Proposition 7.17. It is straightforward to show that if $[p'_i, p'_{i+1}] = [p_{k_1}, p_{k_2}]$ is a valid approximating segment, then any segment $[p_{k_1}, p_k]$, $k_1 < k < k_2$ is also a valid approximating segment. A graph can therefore be constructed using points p_k as vertices linked by arcs representing the fact that an approximating segment between two vertices is valid. The graph thus defined is clearly acyclic (i.e. contains no cycle). When assigning a suitable length to each arc, the shortest path from p_0 to p_n will represent the optimal approximation for the polygonal line $\{p_i\}_{i=0,\ldots,n}$.

In [41, 163], the visibility graph is constructed and the length of each arc is given by the number of original points skipped by an approximating segment $[p'_i, p'_{i+1}] = [p_{k_1}, p_{k_2}]$ ($k_2 - k_1 - 1$, in this case). The minimal configuration is therefore given by the longest path from p_0 to p_n in this graph.

The graph-theoretic approach is now applied in the context defined in [153]. In Section 2.2.1 discrete straightness has been characterised via the chord properties. We will make use of visibility polygons defined by the compact chord property (CCP) for defining a procedure which checks for straightness. The formulation of the compact chord property in the 8-neighbourhood space using visibility polygons is first recalled below.

Proposition 7.18:

An 8-connected set of discrete points $P_{p_0p_n} = \{p_i\}_{i=0,\ldots,n}$ satisfies the compact chord property if and only if, for any two discrete points p_i and p_j in $P_{p_0p_n}$, p_j is strictly visible from p_i in the polygon $O = \{\alpha \in \mathbb{R}^2$ such that $d_4(\alpha, \cup_{k=0,\ldots,n-1}[p_k, p_{k+1}]) < 1\}$.

The construction of the visibility polygons defined by the CCP can be done incrementally while visiting each point p_i. The procedure for the characterisation of the optimal approximation of a chain code sequence is given as the three following steps.

1. Forward visibility analysis:

   With each point $p_i \in P_{p_0p_n}$, we associate an integer value visibility_i such that a point p_j $(j > i)$ is strictly visible from p_i in O if and only if $i < j \leq i + \text{visibility}_i$. Clearly, for $0 \leq i < n$, $1 \leq \text{visibility}_i < n - i$ and $\text{visibility}_n = 0$. The value visibility_i is called the forward visibility of point p_i.

2. Visibility graph construction:

   Using forward visibility values, to each point $p_i \in P_{p_1p_n}$, we associate a point p_j such that $j < i$ and $P_{p_jp_i}$ satisfies the compact chord property. This is the case if p_j is defined as the farthest point from p_i such that p_i is visible from all points p_k such that $j \leq k < i$. In [153], this is referred to as "backward CCP analysis" and allows for defining the set of all possible digital straight segments contained within $P_{p_0p_n}$. More formally, j can be characterised as

   $$j = \min\{m \mid m < i \text{ and } \forall k \text{ such that } m \leq k < i,\ k + \text{visibility}_k \geq i\}$$

   The set of discrete points $P_{p_0p_n}$ is then mapped onto the set of vertices of the visibility graph $G_{\text{vis}} = (V_{\text{vis}}, A_{\text{vis}})$. The set of vertices V_{vis} of G_{vis} are the points of $P_{p_0p_n}$. An arc (p_i, p_j) exists in A_{vis} if and only if $P_{p_ip_j} = \{p_k\}_{k=i,\ldots,j} \subset P_{p_0p_n}$ is a digital straight segment. Based on visibility values, A_{vis} therefore represents the set of all possible digital straight segments contained in $P_{p_0p_n}$.

3. Characterisation of the optimal partition:

   Any arc in A_{vis} is associated with a unit length. The optimal partition of the chain-code sequence $\{c_i\}_{i=1,\ldots,n}$ into a minimal number of digital straight segments reduces to that of locating the set of m vertices traversed by the shortest path from p_0 to p_n in the visibility graph G_{vis}

The above description decomposes the chain code into digital straight segments. However, noisy conditions may occur (particularly when thinning) which make this procedure impractical. Nevertheless, since it is based on a strict characterisation of digital straightness, it readily suggests a relaxation scheme which will allow control of the approximation made in term of the straightness. By simply relaxing the compact chord property, we obtain the following characterisation, based on an approximation threshold $\varepsilon > 1$.

Proposition 7.19:

A connected set of discrete points $P_{p_0 p_n} = \{p_i\}_{i=0,\ldots,n}$ satisfies the relaxed compact chord property for a value $\varepsilon > 1$ if and only if, for any two discrete points p_i and p_j in $P_{p_0 p_n}$, p_j is strictly visible from p_i in the polygon $O = \{\alpha \in \mathbb{R}^2$ such that $d_4(\alpha, \cup_{k=0,\ldots,n-1}[p_k, p_{k+1}]) < \varepsilon\}$.

Only the definition of the visibility value needs to be changed in the relaxed version of the above algorithm. The rest of the procedure is kept unchanged and defines the relaxed polygonal approximation.

The advantage of such an approach is that, in binary line images, the notion of local width can be exploited to calculate an upper bound for ε, thus keeping the approximation step consistent with the rest of the process.

Chapter 8

SOME APPLICATIONS

We conclude this book by presenting some example applications. Five binary images are selected from different application classes. Section 8.1 illustrates the processing of a hand-drawn printed circuit image. The aim here is to convert this image into a vector form. A similar application is presented in Section 8.2, where the input image represents a transportation network. The goal here is essentially to recover the network structure of the image and to measure stroke lengths in order for the final output to give sufficient information for locating the best route between any two points on the processed network. The third image represents a fingerprint (Section 8.3), where identification calls for the characterisation of minutiae within the image. Different classification techniques for fingerprint images have been proposed relying also on the global curvature of ridges within of the fingerprint [93]. We show how thinning techniques presented in Chapter 7 can help in achieving this result. The text image presented in Section 8.4 illustrates the use of skeletonisation for hand-written text analysis. Again, the idea is to define strokes and their inter-relationship for recognition. While all these images can be classified as line images, Section 8.5 illustrates the processing of a generic image. The use of the skeleton for compression in relation to this type of image is discussed in this section.

8.1 Circuit image

In this section, we discuss the successive operations involved in the processing of the image of a printed circuit. We will detail the processing of this image using the graph-theoretic approach presented within this book. The original 573×219 bitmap is shown in Figure 8.1.

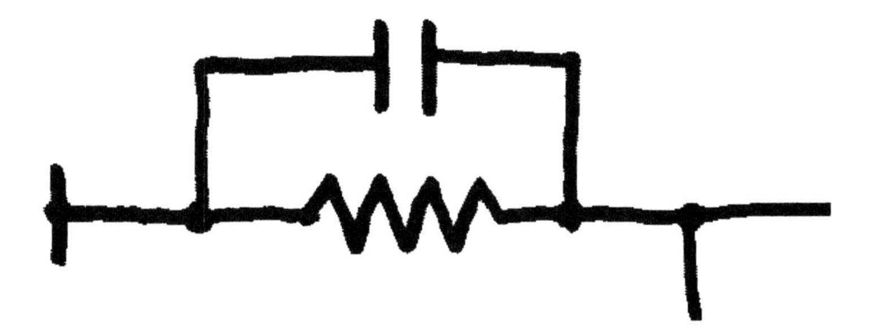

Figure 8.1 Image "Circuit": bitmap

This image is clearly a binary line image, since most of the information is contained within its structure. The first step in our processing will therefore consist of retrieving this upper-level structure using the k-shortest path technique described in Section 7.3.2. The grid graph of this image is first constructed, based on the 8-neighbourhood relationship. This graph contains 13,853 vertices and 51,128 arcs. Finding the upper-level structure of the image characterises a (upper-level) graph with 14 vertices (junctions and end points) connected with 13 arcs (thus characterising a tree). The locations of the upper-level structure vertices are shown as black discs ($\bullet$) in Figure 8.2.

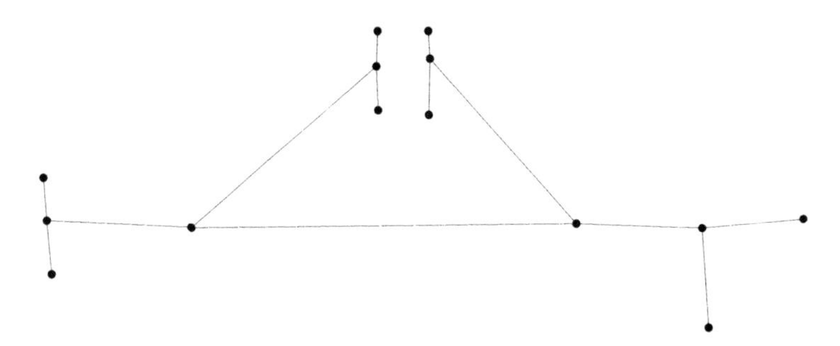

Figure 8.2 Upper-level structure of image "Circuit"

Once the junctions are characterised, they allow for partitioning the image into ribbon-like parts. For each of these parts, we use the graph-theoretic skeletonisation technique based on local digital width [91]. This procedure defines "width lines" within each stroke (see Section 7.2.3). The image in Figure 8.3 shows these width lines when all components are merged back together.

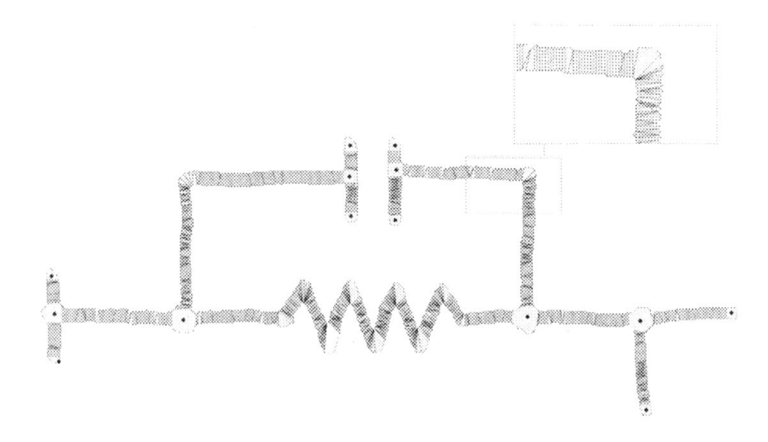

Figure 8.3 Width lines of image "Circuit"

A detail of these width lines which illustrates the behaviour of the procedure at angles is also shown in this figure. Typically, each pixel on the interior border is associated with its closest pixel on the exterior border. The length of these lines will therefore characterise the local width information at each point of the border.

In this context, a weighting scheme associates a centrality weight to each interior point by calculating its distance to the border *along the width line.* This distance may therefore not be equal to the value of the distance transform at this point. The maximum weighted paths joining all pairs of vertices defining the arcs of the upper-level graph now represent the skeleton, as shown in Figure 8.4.

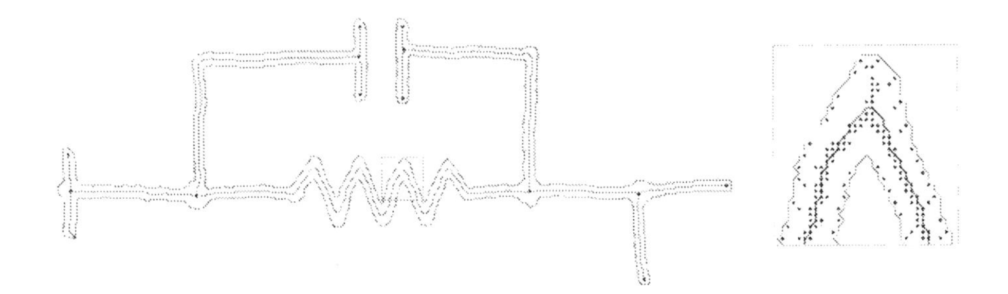

Figure 8.4 Skeleton of image "Circuit"

The fact that the weighting scheme relies on the width information results in a well-located skeleton which does not necessarily follows the set of centres of maximal discs at sharp angles. This is enhanced in Figure 8.4, where black discs are centres of maximal discs within the image and the thick central line is the skeleton based on width information. Therefore, when initiating the vectorisation procedure, the relaxation factor ε can take a lower value than with a skeleton based solely on the distance transform. In our case, Figure 8.5 shows the accuracy of vectorisation for this example (here $\varepsilon = 4$).

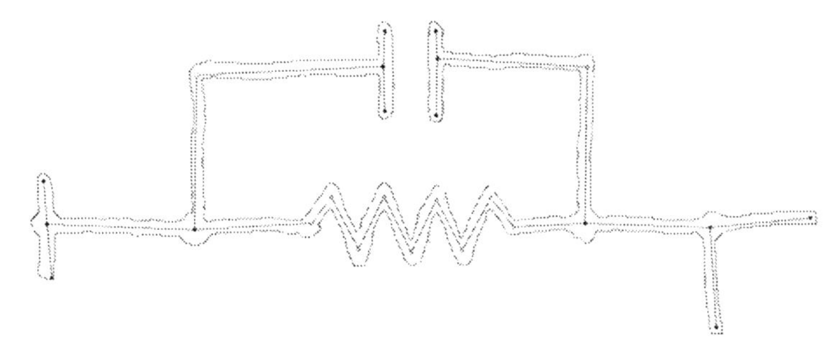

Figure 8.5 Vector form of image "Circuit"

8.2 Road map image

We now show how a similar analysis can be operated on the image of a transportation network. Consider the 230×230 bitmap of the line image presented in Figure 8.6.

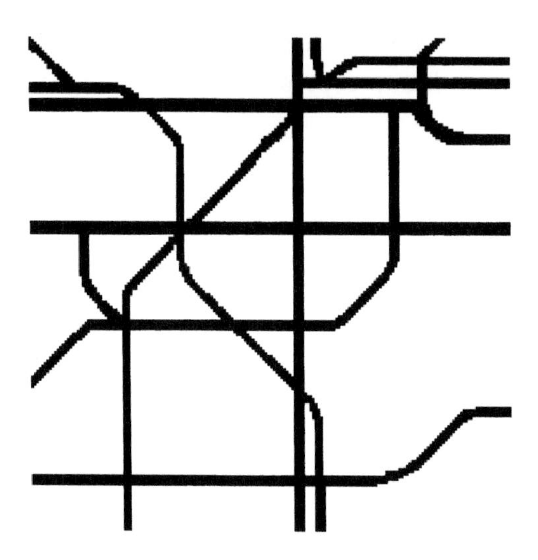

Figure 8.6 Image "Road map": bitmap

The upper -level structure of this image is retrieved as follows. Two points are characterised as extremities of the shortest path of largest length within the grid graph of the image. The set of all possible k-shortest paths between these two points will visit the complete network, excluding branches leading to "free ends". By dismissing these paths and re-iterating the procedure, all central paths within the image are characterised. Their junctions and end points form the vertices of the upper-level graph. Such a graph is shown in Figure 8.7.

For this application, the connectivity between vertices of the upper-level graph is important for finding a route between two given points within the image. It is therefore the structure in which the image will be stored for further analysis. By defining a centrality weight for each arc in the grid graph, the set of maximum weighted paths between the pairs of vertices representing the arcs of the upper-level graph will form the skeleton, as shown in Figure 8.8.

This application highlights the advantage of using graph theory as an approach for defining the skeleton since, at this stage, measuring the length (and other characteristics such as curvature) of each skeleton branch is straightforward. Such characteristics, associated with arcs in the upper-level graph, now form the basis for an automated route planning system.

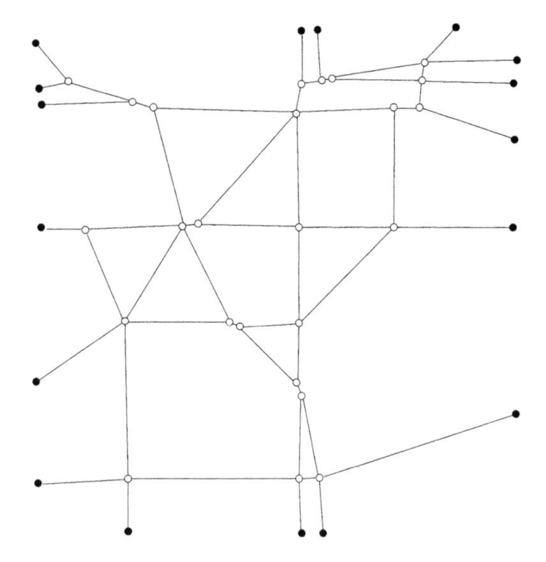

Figure 8.7 Structure of image "Road map"

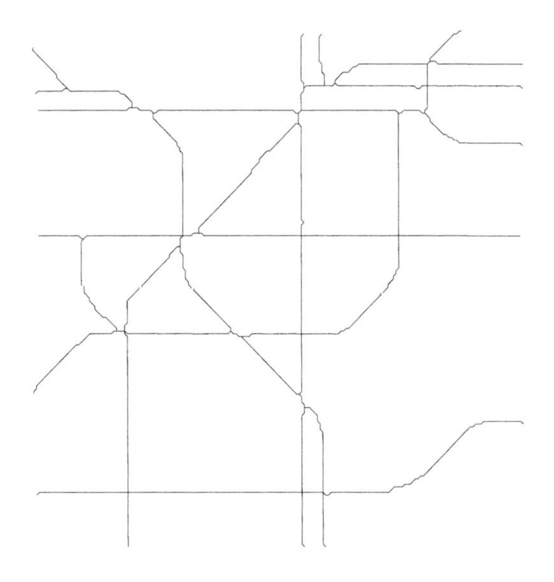

Figure 8.8 Skeleton of image "Road map"

8.3 Fingerprint image

The same thinning technique is applied in view of fingerprint identification. Consider the 315×315 bitmap image shown in Figure 8.9. It is a line image composed of 194 connected components. The construction of the 8-grid graph will result in a graph containing 44,380 vertices and 144,794 arcs and will automatically label all the components. Thinning can therefore be performed in each component separately.

Figure 8.9 Image "Fingerprint": bitmap

Fingerprint identification is typically based on the location of minutiae within the image. Mapping onto our definitions, these features correspond to end and junction vertices within the upper-level graph. Note that, because of the simplicity of each component (mostly a tree structure), this processing is efficiently done by the graph-theoretic procedure. In Figure 8.10, the skeleton and minutiae found are shown. They now form the input for an identification system. Note that, similarly to the previous application, calculating the length and curvature of each skeleton branch is straightforward with the graph structure provided by the thinning procedure.

8.4 Text image

Another major application of line image analysis is hand-written text processing. While typewriter characters are mostly similar in shape from font to font, handwriting may differ substantially from one person to another. The recognition based on the raw bitmap by neural networks, for example, may therefore be difficult in this case. The idea here is to provide an input which is more person-independent than the bitmap itself. In this respect, it is conceivable that the upper-level structure of the text line image may allow the retrieval how the text was actually written and therefore allow identification of the content of the image. Here we are only concerned with the thinning part of the process, which is again considered as a pre-processing stage for recognition.

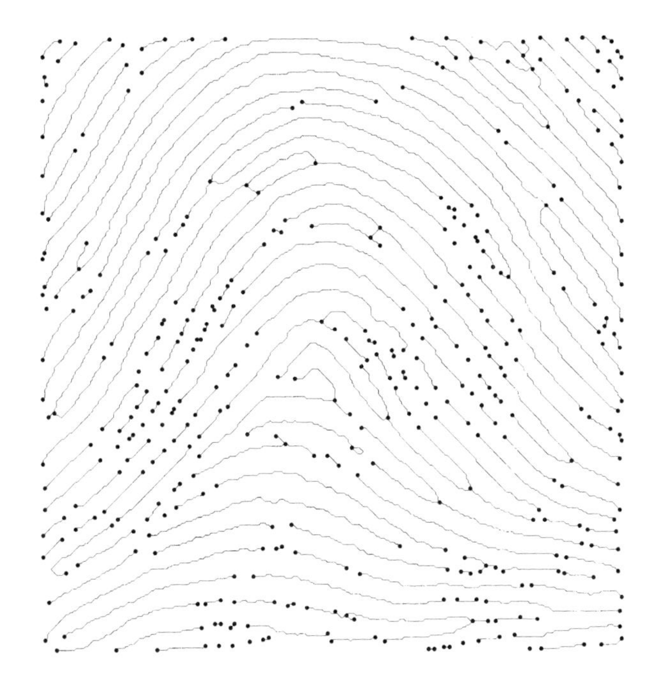

Figure 8.10 Image "Fingerprint": skeleton and minutiae

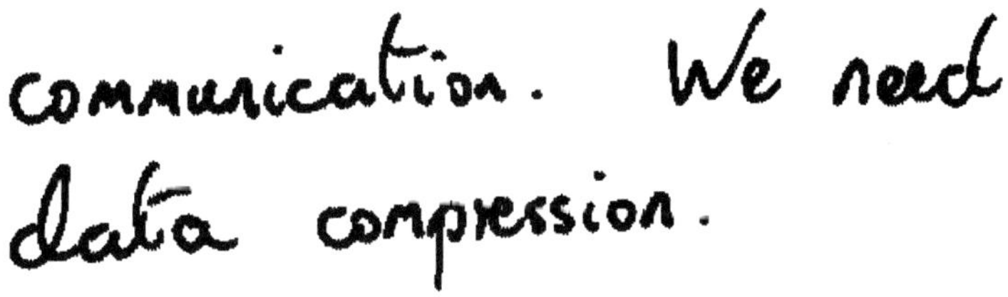

communication. We need data compression.

Figure 8.11 Image "Text": bitmap

Consider the bitmap image shown in Figure 8.11, cropped from a standard fax image. Figure 8.12 shows the location of the centres of maximal discs and the location of the skeleton found within this set for a part of the original image.

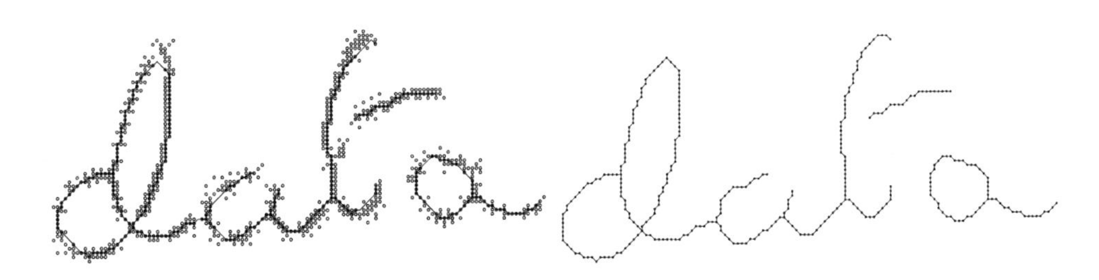

Figure 8.12 Centres of maximal discs and skeleton for a part of image "Text"

For this application, we used the graph-theoretic approach, which ensures connectivity and naturally defines original stroke connectivity. Constructing the grid-graph automatically labels all separated components on which subsequent pre-processing will be applied. Figure 8.13 shows the minimum weighted spanning tree for a simple component of the image. This figure highlights the fact that the skeleton of this component will follow the most central path within this tree (see Figure 8.12).

Figure 8.13 Minimum spanning tree of a part of image "Text"

Finally, all skeleton components are merged together to form the thin input on which features such as junctions and curvatures will be defined for providing an input to a recognition system.

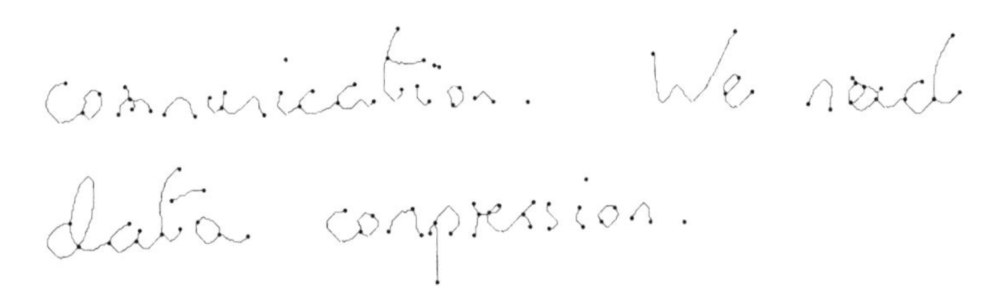

Figure 8.14 Complete skeleton and end points for image "Text"

8.5 Drawing image

While all above applications were based on line images and aimed at identifying features within the image, this section shows how skeletonisation applies to a generic image. The image in question is the 256×256 bitmap image shown in Figure 8.15(A).

Figure 8.15(B) displays the grey-level representation of the $d_{3,4}$-distance map of this image. When locating the centres of maximal discs using the chamfer distance $d_{3,4}$, one finds the set of pixels shown in Figure 8.16(A) (edges are also displayed for clarity). These 1735 pixels allow for a complete and exact reconstruction of the image. All the information is therefore contained in the

Figure 8.15 Image "Squirrel": bitmap and distance map

location of the selected pixels and the corresponding maximal disc radii. In this example, the maximal distance value found is 78, which can be stored in 7 bits. Based on a straightforward calculation, $256^2 = 65,536$ bits are necessary to store the raw bitmap and $(1735 * 8 * 2) + (1735 * 7) = 39,905$ bits are necessary to store the position and radii of the maximal discs. A straightforward 60% cut in the storage capacity needed has therefore been made using this technique.

Figure 8.16 Image "Squirrel": centres of maximal discs and partially reconstructed image

It is clear that more elaborate storage techniques would benefit from this form of the image. On the other hand, all the discs may not be necessary for the complete reconstruction of the image. Some techniques exist which reduce this set while still keeping a lossless representation [52,77,78,88,112,131]. If information loss is accepted, then the set of centres of maximal discs may be decimated on the basis of the corresponding radii, for example, to hierarchically exclude thin details of the image while obtaining a compact storage. In Figure 8.16(B), only the 1220 centres of maximal discs whose corresponding radii exceed 9 are kept. The image is not modified in a dramatic fashion and the storage has now moved to $(1220 * 8 * 2) + (1220 * 7) = 28,060$ bits.

Figure 8.17 Image "Squirrel": skeleton and skeleton subtracted from the original

Figure 8.17(A) finally shows the skeleton of this image obtained using a morphological thinning technique. The skeleton is also shown as subtracted from the original image to highlight how each skeleton branch hierarchically represents a sub-part of the image (Figure 8.17(B)). This also shows the correspondence between the skeleton and the ridges of the distance map of the image shown in Figure 8.15(B).

REFERENCES

[1] C. Arcelli and G. Sanniti di Baja. Computing Voronoi diagrams in digital pictures. *Pattern Recognition Letters*, 4(5):383–389, 1986.

[2] C. Arcelli and G. Sanniti di Baja. Finding local maxima in a pseudo-Euclidean distance transform. *Computer Vision, Graphics and Image Processing*, 43:361–367, 1988.

[3] C. Arcelli and G. Sanniti di Baja. Picture editing by simultaneously smoothing figure protusions and dents. In *Proceedings of the Ninth International Conference on Pattern Recognition*, pages 948–950, Rome, Italy, November 1988.

[4] A. L. D. Beckers and A. W. M. Smeulders. A comment on "A note on 'Distance transformations in digital images' ". *Computer Vision, Graphics and Image Processing*, 47:89–91, 1989.

[5] C. A. Berenstein and D. Lavine. On the number of digital straight line segments. *IEEE Transactions on Pattern Analysis and Machine Intelligence*, PAMI-10(6):880–887, 1988.

[6] P. Bhattacharya. Connected component labeling for binary images on a reconfigurable mesh architecture. *Journal of Systems Architecture*, 42(4):309–313, 1996.

[7] S. N. Biswas and B. B. Chaudhuri. On the generation of discrete circular objects and their properties. *Computer Vision, Graphics and Image Processing*, 32:158–170, 1985.

[8] H. Blum. A transformation for extracting new descriptors of shape. In W. Wathen-Dunn, editor, *Models for the Perception of Speech and Visual Form*, pages 362–380. MIT Press, Cambridge Mass., 1967.

[9] H. Blum. Biological shape and visual science (Part 1). *Journal of Theoretical Biology*, 38:205–287, 1973.

[10] H. Blum and R. N. Nagel. Shape description using weighted symmetric axis features. *Pattern Recognition*, 10:167–180, 1978.

[11] A. Bogomolny. Digital geometry may not be discrete. *Computer Vision, Graphics and Image Processing*, 43:205–220, 1988.

[12] G. Borgefors. Distance transformations in arbitrary dimensions. *Computer Vision, Graphics and Image Processing*, 27:321–345, 1984.

[13] G. Borgefors. Distance transformations in digital images. *Computer Vision, Graphics and Image Processing*, 34:344–371, 1986.

[14] G. Borgefors. Distance transformations on hexagonal grids. *Pattern Recognition Letters*, 9(2):97–105, 1989.

[15] G. Borgefors. Another comment on "A note on 'Distance transformations in digital images' ". *CVGIP: Image Understanding*, 54(2):301–306, 1991.

[16] J. E. Bresenham. Algorithm for computer control of a digital plotter. *IBM Systems Journal*, 4(1):25–30, 1965.

[17] J. E. Bresenham. A linear algorithm for incremental digital display of circular arcs. *Computer Graphics and Image Processing*, 20(2):100–106, 1977.

[18] H. Breu, J. Gil, D. Kirkpatrick and M. Werman. Linear-time Euclidean distance transform algorithms. *IEEE Transactions on Pattern Analysis and Machine Intelligence*, 17(5):529–533, 1995.

[19] S.-K. Chang. *Principles of pictorial information systems design.* Prentice-Hall, Englewood Cliffs, N.J., 1989.

[20] J.-M. Chassery. Discrete convexity: definitions, parametrization and compatibility with continuous convexity. *Computer Vision, Graphics and Image Processing*, 21:326–344, 1983.

[21] J.-M. Chassery and M. I. Chenin. Topologies on discrete spaces. In Simon & Haralick, editor, *Digital Image Processing*, pages 59–66. Reidel, 1980.

[22] J.-M. Chassery and A. Montanvert. *Géométrie Discrète en Analyse d'Images.* Editions Hermès, Paris, 1991. In French.

[23] S. Chattopadhyay and P. P. Das. A new method of analysis for discrete straight lines. *Pattern Recognition Letters*, 12:747–755, 1991.

[24] L. Chen and H. Y. H. Chuang. A fast algorithm for Euclidean distance maps of a 2-D binary image. *Information Processing Letters*, 51:25–29, 1994.

[25] N. Christofides. *Graph Theory: An Algorithmic Approach.* Academic Press, London, 1975.

[26] N. Christofides, H. O. Badra and Y. M. Sharaiha. Data structures for topological and geometric operations on networks. *Annals of Operations Research: OR/IS Interface*, 71:259–289, 1997.

[27] P. Clermont and B. Zavidovique. Communication control in a pyramid computer: application to image region labeling. In *Proceedings of the Tenth International Conference on Pattern Recognition*, pages 551–555, Atlantic City, June 1990. IEEE Computer Society Press.

[28] P.-E. Danielsson. Euclidean distance mapping. *Computer Graphics and Image Processing*, 14:227–248, 1980.

[29] P. P. Das. An algorithm for computing the number of minimal paths in digital images. *Pattern Recognition Letters*, 9:107–116, 1989.

[30] P. P. Das and B. N. Chatterji. Knight's distances in digital geometry. *Pattern Recognition Letters*, 7:215–226, 1988.

[31] P. P. Das and J. Mukherjee. Metricity of super knight's distance in digital geometry. *Pattern Recognition Letters*, 11:601–604, 1990.

[32] S. di Zenzo, S. Cinque and S. Levialdi. Run-based algorithms for binary image-analysis and processing. *IEEE Transactions on Pattern Analysis and Machine Intelligence*, PAMI-18(1):83–89, 1996.

[33] R. B. Dial. Algorithm 360: Shortest path forest with topological ordering. *Communications of the ACM*, 12(11):632–633, 1969.

[34] R. B. Dial, F. Glover, D. Karney and D. Klingman. A computational analysis of alternative algorithms and labelling techniques for finding shortest paths trees. *Networks*, 9:215–248, 1979.

[35] E. W. Dijkstra. A note on two problems in connexion with graphs. *Numerical Mathematics*, 1:269–271, 1959.

[36] L. Dorst and R. P. W. Duin. Spirograph theory: a framework for calculations on digitized straight lines. *IEEE Transactions on Pattern Analysis and Machine Intelligence*, PAMI-6(5):632–639, 1984.

[37] L. Dorst and A. W. M. Smeulders. Discrete representation of straight lines. *IEEE Transactions on Pattern Analysis and Machine Intelligence*, PAMI-6(4):450–463, 1984.

[38] L. Dorst and P. W. Verbeek. The constrained distance transformation: A pseudo-Euclidean, recursive implementation of the Lee-algorithm. In *European Signal Processing Conference (EUSIPCO-86)*, pages 917–920, 1986.

[39] E. R. Dougherty and C. R. Giardina. *Image Processing – Continuous to Discrete*, volume I. Prentice-Hall, Englewood Cliffs, N. J., 1987.

[40] H. Edelsbrunner. *Algorithms in Computational Geometry.* Springer-Verlag, New York, 1988.

[41] D. Eu and G. T. Toussaint. On approximating polygonal curves in two or three dimensions. *CVGIP: Graphical Models and Image Processing,* 56(3):231–246, 1994.

[42] S. Forchhammer. Euclidean distances from chamfer distances for limited distances. In *Proceedings of the Sixth Scandinavian Conference on Image Analysis,* pages 393–400, Oulu, Finland, June 19–22 1989.

[43] H. Freeman. Boundary encoding and processing. In B. S. Lipkin and A. Rosenfeld, editors, *Picture Processing and Psychopictorics,* pages 241–266. Academic Press, New York, 1970.

[44] H. Freeman. Computer processing of line-drawing images. *Computing Surveys,* 6(1):57–97, 1974.

[45] H. Freeman. Algorithm for generating a digital straight line on a triangular grid. *IEEE Transactions on Computers,* C-28(2):150–152, 1979.

[46] A. Fujiwara, T. Masuzawa and H. Fujiwara. An optimal parallel algorithm for the Euclidean distance maps of 2-D binary images. *Information Processing Letters,* 54:295–300, 1995.

[47] G. Gallo and S. Pallottino. Shortest path algorithms. *Annals of Operations Research,* 13:3–79, 1988.

[48] C. R. Giardina and E. R. Dougherty. *Morphological Methods in Image and Signal Processing.* Prentice-Hall, Englewood Cliffs, N. J., 1988.

[49] M. Gondran and M. Minoux. *Graphs and Algorithms.* Wiley-Interscience Series in Discrete Mathematics. John Wiley and Sons, 1984.

[50] R. C. Gonzalez and R. E. Woods. *Digital Image Processing.* Addison Wesley, Reading, Mass., 1992. Reprinted with corrections September, 1993.

[51] J. E. Goodman, R. Pollack and W. Steiger, editors. *Discrete and computational geometry: papers from the DIMACS special year.* American Mathematical Society – Association for Computing Machinery (ACM), 1991.

[52] J. Goutsias and D. Schonfeld. Morphological representation of discrete and binary images. *IEEE Transactions on Signal Processing,* 39(6):1369–1379, June 1991.

[53] H. P. A. Haas. *Convexity analysis of hexagonally sampled images.* PhD thesis, Technishe Hogeschool Delft, Delft, The Netherlands, 1985.

[54] F. Harary, R. A. Melter and I. Tomescu. Digital metrics: a graph-theoretical approach. *Pattern Recognition Letters*, 2:159–163, 1984.

[55] P. Hart, N. Nilsson and B. Raphael. A formal basis for the heuristic determination of minimum cost paths. *IEEE Transactions on Systems, Science and Cybernetics*, SSC-4(2):100–107, 1968.

[56] C. J. Hilditch. Linear skeletons from square cupboards. In B. Meltzer and D. Mitchie, editors, *Machine Intelligence*, volume 4, pages 403–420. Edinburgh University Press, Edinburgh, 1969.

[57] C. J. Hilditch and D. Rutovitz. Chromosome recognition. *Annals of the New York Academy of Sciences*, 157:339–364, 1969.

[58] H. Holin. Harthong-Reeb analysis and digital circles. *Visual Computer*, 8(1):8–17, 1991.

[59] H. Holin. Some artifacts of integer-computed circles. *Annals of Mathematics and Artificial Intelligence*, 16(1–4):153–181, 1996.

[60] C. T. Huang and O. R. Mitchell. A Euclidean distance transform using grayscale morphology decomposition. *IEEE Transactions on Pattern Analysis and Machine Intelligence*, PAMI-16(4):443–448, 1994.

[61] S. H. Y. Hung. On the straightness of digital arcs. *IEEE Transactions on Pattern Analysis and Machine Intelligence*, PAMI-7(2):203–215, 1985.

[62] A. K. Jain. *Fundamentals of Digital Image Processing*. Prentice-Hall Information and System Sciences Series. Prentice-Hall International, London, 1989.

[63] T. Kato, T. Hirata, T. Saito and K. Kise. An efficient algorithm for the Euclidean distance transformation. *Systems and Computers in Japan*, 27(7):18–24, 1996.

[64] E. Khalimsky, R. Kopperman and P. R. Meyer. Computer graphics and connected topologies on finite ordered sets. *Topology and Its Applications*, 36(1):1–17, 1990.

[65] C. E. Kim. On the cellular convexity of complexes. *IEEE Transactions on Pattern Analysis and Machine Intelligence*, PAMI-3:617–625, 1981.

[66] C. E. Kim. Digital convexity, straightness and convex polygons. *IEEE Transactions on Pattern Analysis and Machine Intelligence*, PAMI-4:618–626, 1982.

[67] C. E. Kim. On cellular straight line segments. *Computer Graphics and Image Processing*, 18:369–381, 1982.

[68] C. E. Kim. Digital disks. *IEEE Transactions on Pattern Analysis and Machine Intelligence*, PAMI-6(3):372–374, 1984.

[69] C. E. Kim and A. Rosenfeld. Digital straight lines and convexity of digital regions. *IEEE Transactions on Pattern Analysis and Machine Intelligence*, PAMI-4(2):149–153, 1982.

[70] C. E. Kim and J. Sklansky. Digital and cellular convexity. *Pattern Recognition*, 15:359–367, 1982.

[71] R. Klette, A. Rosenfeld and F. Slodoba, editors. *Advances in Digital and Computational Geometry.* Springer-Verlag, Singapore, 1998.

[72] M. N. Kolountzakis and K. N. Kutulakos. Fast computation of Euclidean distance maps for binary images. *Information Processing Letters*, 43:181–184, 1992.

[73] T. Y. Kong and A. Rosenfeld, editors. *Journal of Mathematical Imaging and Vision: Special Issue on Topology and Geometry in Computer Vision*, volume 6. Kluwer Academic Publishers, 1996.

[74] T. Y. Kong and A. Rosenfeld, editors. *Topological Algorithms for Digital Image Processing.* North-Holland, 1996.

[75] J. Koplowitz and A. M. Bruckstein. Design of perimeter estimators for digitized planar shapes. *IEEE Transactions on Pattern Analysis and Machine Intelligence*, PAMI-11(6):611–622, 1989.

[76] J. Koplowitz, M. Lindenbaum and A. M. Bruckstein. The number of digital straight-lines on a $n \times n$ grid. *IEEE Transactions on Information Theory*, IT-36:192–197, 1990.

[77] D. Kresch and D. Malah. Morphological reduction of skeleton redundancy. *Signal Processing*, 38(1):143–151, 1994.

[78] R. Kresch and D. Malah. Skeleton-based morphological coding of binary images. *IEEE Transactions on Image Processing*, IP-7(10):1387–1394, 1998.

[79] R. Krishnaswamy and C. E. Kim. Digital parallelism, perpendicularity and rectangles. *IEEE Transactions on Pattern Analysis and Machine Intelligence*, PAMI-9(2):316–321, 1987.

[80] J. B. Kruskal. On the shortest spanning subtree of a graph and the travelling salesman problem. *Proceedings of the American Mathematical Society*, 71:48–50, 1956.

[81] Z. Kulpa and B. Kruse. Algorithms for circular propagation in discrete images. *Computer Vision, Graphics and Image Processing*, 24:305–328, 1983.

[82] L. Lam, S. W. Lee and C. Y. Suen. Thinning methodologies – A comprehensive survey. *IEEE Transactions on Pattern Analysis and Machine Intelligence*, PAMI-14(9):869–885, 1992.

[83] D. T. Lee and F. P. Preparata. Euclidean shortest paths in the presence of rectilinear barriers. *Networks*, 14:393–410, 1984.

[84] T. Lindeberg. *Scale-Space Theory in Computer Vision.* Kluwer Academic Publishers, Dordrecht, 1994.

[85] M. Lindenbaum and J. Koplowitz. A new parametrisation of digital straight lines. *IEEE Transactions on Pattern Analysis and Machine Intelligence*, PAMI-13:847–852, 1991.

[86] D. G. Lowe. Three-dimensional object recognition from single two-dimensional images. *Artificial Intelligence*, 31(3):355–395, 1987.

[87] P. Maragos. Morphological systems: Slope transforms and min-max difference and differential equations. *Signal Processing*, 38:57–77, 1994.

[88] P. Maragos and R. W. Schafer. Morphological skeleton representation and coding of binary images. *IEEE Transactions on Acoustics, Speech, and Signal Processing*, ASSP-34(5):1228–1244, 1986.

[89] S. Marchand-Maillet and Y. M. Sharaiha. Discrete convexity, straightness and the 16-neighbourhood. *Computer Vision and Image Understanding*, 66(3):316–329, 1997.

[90] S. Marchand-Maillet and Y. M. Sharaiha. A graph-theoretic algorithm for the exact solution of the Euclidean distance mapping. In *Tenth Scandinavian Conference on Image Analysis*, volume I, pages 221–228, Lappeenranta, Finland, June 9–11 1997.

[91] S. Marchand-Maillet and Y. M. Sharaiha. Skeleton location and evaluation based on local digital width in ribbon-like images. *Pattern Recognition*, 30(11):1855–1865, 1997.

[92] S. Marchand-Maillet and Y. M. Sharaiha. Euclidean ordering via chamfer distance calculations. *Computer Vision and Image Understanding*, 73(3):404–413, 1999.

[93] K. V. Mardia, Q. Li and T. J. Hainsworth. On the Penrose hypothesis of fingerprint patterns. *IMA Journal of Mathematics Applied in Medicine and Biology*, 9(4):289–294, 1992.

[94] R. A. Melter. New views of linearity and connectedness in digital geometry. *Pattern Recognition Letters*, 10:9–16, 1989.

[95] M. L. Minsky and S. A. Papert. *Perceptrons: An Introduction to Computational Geometry.* M.I.T. Press, Cambridge Mass., expanded edition, 1988.

[96] K. Miyaoku and K. Harada. Approximating polygonal curves in two and three dimensions. *Graphical Models and Image Processing*, 60(3):222–225, 1998.

[97] U. Montanari. A method for obtaining skeletons using a quasi-Euclidean distance. *Journal of the ACM*, 15(4):600–624, 1968.

[98] U. Montanari. Continuous skeletons from digitized images. *Journal of the ACM*, 16:534–549, 1969.

[99] U. Montanari. A note on minimal length polygonal approximation to a digitized contour. *Communications of the ACM*, 13:41–47, 1970.

[100] A. Montanvert. Medial line: Graph representation and shape description. In *Eighth International Conference on Pattern Recognition*, pages 430–432, Paris, France, October 27–31 1986. IEEE Computer Society Press, Washington D.C.

[101] E. F. Moore. The shortest path through a maze. In *International Symposium on the Theory of Switching Proceedings, Part II*, pages 285–292. Harvard U. Press, Cambridge, Mass., 1959.

[102] O. J. Morris, M. de Jersey Lee and A. G. Constantinides. Graph theory for image analysis: An approach based on the shortest spanning tree. *IEE Proceedings-F Communications Radar and Signal Processing*, 133(2):146–152, 1986.

[103] P. F. M. Nacken. Chamfer metrics, the Medial Axis and mathematical morphology. *Journal of Mathematical Imaging and Vision*, 6:235–248, 1996.

[104] A. Nakamura and K. Aizawa. Digital circles. *Computer Vision, Graphics and Image Processing*, 26(2):242–255, 1984.

[105] A. Nakamura and K. Aizawa. Digital images of geometric pictures. *Computer Vision, Graphics and Image Processing*, 30:107–120, 1985.

[106] A. Nakamura and K. Aizawa. Digital squares. *Computer Vision, Graphics and Image Processing*, 49(3):357–368, 1990.

[107] K. Nakashima, M. Koga, K. Marukawa, Y. Shima and Y. Nakano. A high-speed contour fill method for character image generation. *IEICE Transactions on Information and Systems*, E77D(7):832–838, 1994.

[108] G. Nemhauser. A generalized permanent label setting algorithm for the shortest path between two specified nodes. *Journal of Mathematical Analysis and Applications*, 38:328–334, 1972.

[109] A. N. Netravali and B. G. Haskell. *Digital Pictures: Representation, Compression, and Standards.* Plenum Press, New York, second edition, 1994.

[110] S. Olariu, J. L. Schwing and J. Y. Zhang. Fast component labeling and convex-hull computation on reconfigurable meshes. *Image and Vision Computing*, 11(7):447–455, 1993.

[111] D. W. Pagliero. Distance transforms. *CVGIP: Graphical Models and Image Processing*, 54:56–74, 1992.

[112] T. W. Pai and J. H. L. Hansen. Boundary-constrained morphological skeleton minimization and skeleton reconstruction. *IEEE Transactions on Pattern Analysis and Machine Intelligence*, PAMI-16(2):201–208, 1994.

[113] U. Pape. Implementation and efficiency of Moore algorithms for the shortest route problem. *Mathematical Programming*, 7:212, 1974.

[114] U. Pape. Algorithm 562: shortest path lengths. *ACM Transactions on Mathematical Software*, 6(3):450–455, 1980.

[115] U. Pape. Remark on algorithm 562. *ACM Transactions on Mathematical Software*, 9(2):260, 1983.

[116] T. Pavlidis. *Algorithms for Graphics and Image Processing.* Computer Science Press, Rockville, Md., 1982.

[117] W. B. Pennebaker and J. L. Mitchell. *JPEG Still Image Compression Standard.* Van Nostrand Reinhold, New York, 1993.

[118] S. Pham. Digital straight segments. *Computer Vision, Graphics and Image Processing*, 36:10–30, 1986.

[119] J. Piper and E. Granum. Computing distance transformations in convex and non-convex domains. *Pattern Recognition*, 20(6):599–615, 1987.

[120] F. P. Preparata and M. I. S. Shamos. *Computational Geometry, an Introduction.* Springer-Verlag, New York, 1985.

[121] R. C. Prim. Shortest connection networks and some generalizations. *Bell System Technical Journal*, 36:1389, 1957.

[122] K. Qian and P. Bhattacharya. Determining holes and connectivity in binary images. *Computers and Graphics*, 16(3):283–288, 1992.

[123] I. Ragnemalm. Contour processing distance transforms. In Cantoni *et al.*, editor, *Progress in Image Analysis and Processing*, pages 204–212. World Scientific, 1990.

[124] I. Ragnemalm. *Generation of Euclidean Distance Maps*. Linköping Studies in Science and Technology. Linköping University, Linköping, Sweden, 1990. Licenciate Thesis No. 206.

[125] I. Ragnemalm. Fast edge smoothing in binary images using Euclidean metric. In Cantoni *et al.*, editor, *Progress in Image Analysis II*. World Scientific, Singapore, 1992.

[126] I. Ragnemalm. Fast erosion and dilation by contour processing and thresholding of distance maps. *Pattern Recognition Letters*, 13(3):161–166, 1992.

[127] I. Ragnemalm. Neighborhoods for distance transformations using ordered propagations. *CVGIP: Image Understanding*, 56:399–409, 1992.

[128] I. Ragnemalm. *The Euclidean Distance Transform*. Linköping Studies in Science and Technology. Linköping University, Linköping, Sweden, 1993. Dissertation No. 304.

[129] I. Ragnemalm. The Euclidean distance transform in arbitrary dimensions. *Pattern Recognition Letters*, 14(11):883–888, 1993.

[130] I. Ragnemalm. A note on "Optimization on Euclidean distance transformation using grayscale morphology". In *The Euclidean Distance Transform*, pages 263–276. Linköping University, Linköping, Sweden, 1993.

[131] J. M. Reinhardt and W. E. Higgins. Efficient morphological shape representation. *IEEE Transactions on Image Processing*, 5(1):89–101, 1996.

[132] C. Ronse. Definitions of convexity and convex hulls in digital images. *Bull. Soc. Math. Belge Série B*, 37(2):71–85, 1985.

[133] C. Ronse. An isomorphism for digital images. *Journal of Combinatorial Theory, Series A*, 39:132–159, 1985.

[134] C. Ronse. A simple proof of Rosenfeld's characterisation of digital straight segments. *Pattern Recognition Letters*, 3(5):323–326, 1985.

[135] C. Ronse. A strong chord property for 4-connected convex digital sets. *Computer Vision, Graphics and Image Processing*, 35:259–269, 1986.

[136] C. Ronse. A bibliography on digital and computational convexity (1961–1988). *IEEE Transactions on Pattern Analysis and Machine Intelligence*, PAMI-11(2):181–189, 1989.

[137] A. Rosenfeld. Digital straight line segments. *IEEE Transactions on Computers*, C-23(12):1264–1269, 1974.

[138] A. Rosenfeld. Digital topology. *American Mathematical Monthly*, 86:621–629, 1979.

[139] A. Rosenfeld and E. Johnston. Angle detection on digital curves. *IEEE Transactions on Computers*, pages 875–878, 1973.

[140] A. Rosenfeld and C. E. Kim. How a digital computer can tell us whether a line is straight. *American Mathematical Monthly*, 89:230–235, 1982.

[141] A. Rosenfeld and J. L. Pfaltz. Sequential operations in digital picture processing. *Journal of the ACM*, 13(4):471–497, 1966.

[142] A. Rosenfeld and J. L. Pfaltz. Distances functions on digital pictures. *Pattern Recognition*, 1:33–61, 1968.

[143] P. L. Rosin. Techniques for assessing polygonal approximations of curves. *IEEE Transactions on Pattern Analysis and Machine Intelligence*, 19(6):659–666, 1997.

[144] P. L. Rosin and G. A. W. West. Segmentation of edges into lines and arcs. *Image and Vision Computing*, 7(2):109–114, 1989.

[145] H. Samet. The quadtree and related hierarchical data structures. *Computing Surveys*, 16(2):187–260, 1984.

[146] H. Samet. *Applications of Spatial Data Structures.* Addison Wesley, Reading, Mass., 1990.

[147] H. Samet. *The Design and Analysis of Spatial Data Structures.* Addison Wesley, Reading, Mass., 1990.

[148] P. V. Sankar and E. V. Krishnamurty. On the compactness of subsets of digital pictures. *Computer Graphics and Image Processing*, 8:136–143, 1978.

[149] J. Serra. *Image Analysis and Mathematical Morphology.* Academic Press, London, 1982.

[150] J. Serra. *Image Analysis and Mathematical Morphology, Part II: Theoretical Advances.* Academic Press, London, 1988.

[151] B. Shapiro, J. Pisa and J. Sklansky. Skeleton generation from x,y boundary sequences. *Computer Graphics and Image Processing*, 15:136–153, 1981.

[152] Y. M. Sharaiha. *A graph theoretic approach for the raster-to-vector problem in digital image processing.* PhD thesis, Imperial College, London, 1991.

[153] Y. M. Sharaiha and N. Christofides. An optimal algorithm for straight segment approximation of digital arcs. *CVGIP: Graphical Models and Image Processing*, 55(5):397–407, 1993.

[154] Y. M. Sharaiha and N. Christofides. A graph theoretic approach to distance transformations. *Pattern Recognition Letters*, 15(10):1035–1041, 1994.

[155] Y. M. Sharaiha and P. Garat. A compact chord property for digital arcs. *Pattern Recognition*, 26(5):799–803, 1993.

[156] Y. M. Sharaiha and R. Thaiss. Guided search for the shortest path on transportation networks. In I. H. Osman and J. P. Kelly, editors, *Meta-Heuristics: Theory and Applications*, pages 115–132. Kluwer Academic Publishers, 1996.

[157] F. Y. Shih and H. Wu. Optimization on Euclidean distance transformation using gray-scale morphology. *Journal of Visual Communication and Image Representation*, 3:104–114, 1992.

[158] F. Y. C. Shih and O. R. Mitchell. A mathematical morphology approach to Euclidean distance transformation. *IEEE Transactions on Image Processing*, 1(2):197–204, 1992.

[159] J. Sklansky. Recognition of convex blobs. *Pattern Recognition*, 2:3–10, 1970.

[160] J. Stoer and C. Witzgall. *Convexity and Optimization in Finite Dimensions I.* Springer-Verlag, 1970.

[161] S. Suzuki, N. Ueda and J. Sklansky. Graph-based thinning for binary images. *International Journal of Pattern Recognition and Artificial Intelligence*, 7(5):1009–1030, 1993.

[162] E. Thiel and A. Montanvert. Chamfer masks: discrete distance functions, geometrical properties and optimization. In *Eleventh International Conference on Pattern Recognition*, pages 244–247, The Hague, The Netherlands, August 30–September 3 1992.

[163] G. T. Toussaint. On the complexity of approximating polygonal curves in the plane. In *Proceedings, IASTED International Symposium on Robotics and Automation*, pages 59–62, Lugano, Switzerland, 1985.

[164] L. J. van Vliet and B. J. H. Verwer. A contour processing method for fast binary neighborhood operations. *Pattern Recognition Letters*, 7:27–36, 1988.

[165] B. J. H. Verwer, P. W. Verbeek and S. T. Dekker. An efficient uniform cost algorithm applied to distance transforms. *IEEE Transactions on Pattern Analysis and Machine Intelligence*, 11(4):425–429, 1989.

[166] J. H. Verwer. Local distance for distance transformations in two and three dimensions. *Pattern Recognition Letters*, 12:671–682, 1991.

[167] K. Voss. *Discrete images, objects and functions in* $\mathbb{Z}^n$. Algorithms and Combinatorics, 11. Springer-Verlag, Berlin, 1993.

[168] A. M. Vossepoel. A note on "Distance transformations in digital images". *Computer Vision, Graphics and Image Processing*, 43:88–97, 1988.

[169] A. M. Vossepoel and A. W. M. Smeulders. Vector code probability and metrification error in the representation of straight lines of finite length. *Computer Graphics and Image Processing*, 20:347–364, 1982.

[170] G. A. W. West and P. L. Rosin. Techniques for segmenting image curves into meanigful descriptions. *Pattern Recognition*, 24(7):643–652, 1991.

[171] L. D. Wu. On the chain-code of a line. *IEEE Transactions on Pattern Analysis and Machine Intelligence*, PAMI-4:347–353, 1982.

[172] X. L. Wu and J. G. Rokne. Double-step incremental generation of lines and circles. *Computer Vision, Graphics and Image Processing*, 37(3):331–344, 1987.

[173] H. Yamada. Complete Euclidean distance transformation by parallel operation. In *Proceedings of the Seventh International Conference on Pattern Recognition*, volume 1, pages 69–71, Montreal, Canada, 1984.

[174] M. Yamashita and N. Honda. Distance functions defined by variable neighbourhood sequences. *Pattern Recognition*, 17(5):509–513, 1984.

[175] M. Yamashita and T. Ibaraki. Distances defined by neighbourhood sequences. *Pattern Recognition*, 19(3):237–246, 1986.

[176] Q. Z. Ye. The signed Euclidean distance transform and its applications. In *Proceedings of the Ninth International Conference on Pattern Recognition*, pages 495–499, Rome, Italy, November 1988.

[177] Q. Z. Ye. A fast algorithm for convex-hull extraction in 2D images. *Pattern Recognition Letters*, 16(5):531–537, 1995.

[178] S. Yokoi, J.-I. Toriwaki and T. Fukumura. On generalized distance transformation of digitized pictures. *IEEE Transactions on Pattern Analysis and Machine Intelligence*, PAMI-3(4):424–443, 1981.

[179] D. Yu and H. Yan. An efficient algorithm for smoothing, linearization and detection of structural feature points of binary image contours. *Pattern Recognition*, 30(1):57–69, 1997.

Index

Printed and bound by CPI Group (UK) Ltd, Croydon, CR0 4YY
08/05/2025
01864907-0005